THE ASTROLOGICAL HISTORY OF MĀSHĀ'ALLĀH

Harvard Monographs
in the History of Science

THE ASTROLOGICAL HISTORY OF MĀSHĀ'ALLĀH

E. S. KENNEDY
AND
DAVID PINGREE

Harvard University Press
Cambridge, Massachusetts
1971

Distributed in Great Britain by Oxford University Press

Library of Congress Catalog Card Number 72-113185

SBN 674-05025-8

Printed in the United States of America

PREFACE

This study is an attempt to assemble and analyze all that is presently available of a major work by an eighth-century Jewish astrologer known in medieval Europe as Messahala and in the Near East as Māshā'allāh. The results contribute to an understanding of the origin and transmission of astrological doctrine. Beyond this, however, since the astronomical computations upon which Māshā'allāh based his predictions are known to have been made by the use of one of the royal astronomical canons promulgated by the Sasanian dynasty of Iran, it has been possible to infer from his work a good deal of information concerning Sasanian astronomy. No pre-Islamic Iranian astronomical documents, either in the Pahlavī original or in translation, are extant. Hence any such inferences are materials for the eventual writing of this missing chapter in the history of the exact sciences.

Māshā'allāh's book, an astrological world history called "On Conjunctions, Religions, and Peoples," has not survived intact. A large fragment of it, however, is embedded in a work, itself extant only in parts, by a certain Ibn Hibintā, a Christian astrologer who flourished in Baghdad in the ninth century. This Arabic fragment (Munich Cod. Arab. 852, ff 214v to 233v) supplies most of the material for the present work, and since it contains interpolations by Ibn Hibintā and persons unknown, and since the renderings of certain passages are questionable and a few words remain illegible to us, it is presented here not only in English translation (pp. 39-68), but also in facsimile (pp. 1-38). As presently bound and numbered, a group of folios which should appear in the middle of the fragment have been displaced to the end. In the translation and the commentary these folios have been restored to their correct order, namely, ff. 214v to

215v, then 226r to 233v (234r is blank), then 216r to 225v. However, the folio numbers as they appear on the text itself have been retained for reference. Hence the reader who has occasion to work back and forth between text, translation, and commentary should bear this displacement in mind.

The astrological basis of Māshā'allāh's approach to world history is provided by a Sasanian theory that important religious and political changes are indicated by conjunctions of the planets Saturn and Jupiter, which recur at intervals of about twenty years. Successive conjunctions tend to stay in the same astrological triplicity. After a long time, however, over two centuries, they move along into another triplicity. Any such "shift" of triplicity indicates changes of a more sweeping nature than a simple conjunction – the rise of a new nation or dynasty. The advent of a major prophet, an event most portentous of all, is heralded by the completion of a cycle of shifts through all four triplicities. Predictions are made by casting the horoscope for the instant of the vernal equinox (*taḥwīl al-sana*, year-transfer) of the year in which this conjunction or shift occurs.

The fragment contains sixteen such horoscopes, presumably cast by Māshā'allāh himself, plus a seventeenth which was interpolated later. They are:

Horoscope No.	*Folio of text*	*Date*	*Characterization*	*Conjunction No.*
1	215v	11 Feb. –3380	Shift Indicating the Deluge	121
2	226v	12 Feb. –3360	Conjunction of the Year of the Deluge	122
3	228r	13 Mar. –45	Conjunction Indicating the Christ	289
4	228v	14 Mar. –25	Conjunction for the Nativity of Christ	290
5	229r	14 Mar. –12	Year-Transfer of the Nativity of Christ	
6	230r	19 Mar. 571	Shift Indicating the Rise of Islam	320

7	231v	19. Mar.	630	Conjunction Indicating the Nativity of the Prophet	323
8	232v	19 Mar.	650	Conjunction, Fall of the Sasanians, Rise of the Arabs	324
9	216v	19 Mar.	749	Conjunction, Rise of the ʻAbbasids	329
10	218r	20 Mar.	809	Shift, al-Ma'mūn Becomes Caliph	332
11	219v	20 Mar.	829	Conjunction	333
12	220v	20 Mar.	848	Conjunction	334
13	221v	20 Mar.	868	Conjunction	335
14	222v	20 Mar.	888	Conjunction	336
15	223v	21 Mar.	907	Conjunction	337
16	224v	21 Mar.	928	Conjunction	338
17 (no diagram)	225v	8 Oct.	928	Horoscope of the True Conjunction, cast by Ibn Hibintā. The Rise of the Buwayhids	

In general, the horoscopes are not dated in the text; their dates have been determined from the planetary positions shown in the diagrams. The numerical information accompanying the horoscopes also made additional deductions possible. These are described in a section (pp. 69-88) immediately following the translation. It is shown that Māshā'allāh superposed the conjunction astrology upon a Zoroastrian millennial cosmology in which the duration of the universe is to be 12,000 years. This scheme also involved the notion that the planetary motion commenced in A.D. -5782. The conjunction numbers shown in the last column of the table above have been assigned on this basis. The doctrine also alleges that the Deluge took place in -3360, in contrast to the more usual conjunction astrology, which puts the Flood on 17/18 February -3101, the beginning of the current Kaliyuga in Indian chronology.

Furthermore, it is shown that the version of the royal canon (Pahlavī *zīk ī šahriyārān*, Arabic *zīj al-shāh*) used by

Māshā'allāh was that of the king Kisrā Anūshirwān, having as epoch the year of Kisrā's accession. The mean motions, epoch positions, maximum equations, and method of computing the true positions of the superior planets and the sun are also deduced.

The detailed commentary that follows (pp. 89-125) compares the planetary positions given in the text for each horoscope with the results of computations based on Tuckerman's tables.[1] For all horoscopes the astrological interpretations made by Māshā'allāh are discussed in terms of rules laid down by him in other writings. He died sometime between Conjunctions 332 and 333 (Horoscopes 10 and 11). Hence, for all horoscopes preceding 11, correlations are indicated between actual political events and his "predictions."

The six appendices present material related to the main source. Appendix 1 corroborates the epoch and provenance of Māshā'allāh's tables previously inferred. Appendix 2 gives accession horoscopes computed by Māshā'allāh for the prophet Muḥammad and for all the caliphs through Hārūn al-Rashīd. The positions of Saturn and Jupiter in these configurations verify the essential validity of the conjectured *Zīj al-Shāh* models for the motion of these planets. Appendix 3 is a Latin translation (the only version extant) of Māshā'allāh's book on nativities. Its horoscopes are all shown to have been lifted from other sources, three from the *Pentateuch* of Dorotheus of Sidon (first century). Appendix 4 gives ten more assorted horoscopes attributed to Māshā'allāh, from various sources. The last two appendices quote excerpts from two Latin manuscripts, of the fourteenth and fifteenth centuries respectively, which independently date the shift indicating the Flood at −3380.

This study was made possible by grants from the National Science Foundation to the American University of Beirut, and by the Institute for Advanced Study and the University of Chicago. A number of difficult readings in the Arabic text were elucidated by Professor Jibra'il Jabbur, and Mr. Fuad I. Haddad corrected many errors in the translation. The author-

1. B. Tuckerman, *Planetary, Lunar, and Solar Positions*, 2 vols. (Memoirs, Nos. 56 and 59; American Philosophical Society, Philadelphia, 1962 and 1964).

ities of the Bayerische Staatsbibliothek, Munich, have kindly permitted publication of the text in facsimile. We express gratitude to these institutions and individuals without involving them in responsibility for the shortcomings of our labors. Although the project has been a collaboration throughout, all sections save the translation are essentially the work of Professor Pingree.

E.S.K.

CONTENTS

FOREWORD

Some twenty years ago, Otto Neugebauer, to whom we owe so much for his masterly elucidations of the history of early astronomy and mathematics, eloquently set the case for the value of the study of astrological sources for the history of science. In particular, he urged that a critical and informed attention paid to such "wretched subjects" could, and would, immeasurably assist our understanding of Islamic astronomy and science.

The special relevance that astrology bears in this instance derives, at least in part, from the fact that here we have to do with a conflux of ideas and techniques issuing from Arabic, Syriac, Indian, Iranian, and Greek sources. To untangle and sort out the inputs of these varied sources will bring us a more proper comprehension of the astrological work before us. It will reveal much more of what is involved in such a text, and of the motives behind its composition, than would be possible without such a scholarly excavation and sifting of the sources. The present book does just that. At the same time, it not only evaluates and sets in context Māshā'allāh's astrological world history, but presents a case history of the transmission and dissemination of ideas from other cultures into Islam.

What is more, in uncovering the ideas behind Māshā'allāh's efforts, we are simultaneously informed of much we did not know of these ideas themselves. Thus, merely on the level of the history of astronomy proper, the analysis of the way in which Māshā'allāh has computed the Jupiter-Saturn conjunctions that lie at the heart of his astrological history reveals his dependence upon a previously established millennial theory. This, as well as other factors that the authors have educed from their analysis of the text, has opened a vista, however much a beginning one, on the history of Sasanian astronomy.

But it is not merely the history of astronomy that the present book illuminates. Masha'allah's title for his work was *On Conjunctions, Religions, and Peoples*, a fact that heralds, as it were, its utility for our understanding of other aspects of Islamic culture. Highly technical as the contents of this work might be, it nonetheless also affords us, to turn to Otto Neugebauer once again, "an insight into the daily life, religion and superstition, and astronomical methods and cosmological ideas" of at least a segment of medieval Islam.

John E. Murdoch

THE ASTROLOGICAL HISTORY OF MĀSHĀ'ALLĀH

1. TEXT

214v

بسم الله الرحمن الرحيم رب يسر برحمتك
الحمد لله الحميد المجيد المبدي المعيد الفعال لما يريد خالق
الليل والنهار وعالم ما في البر والبحار وكاشف مكنون الحكم
والاسرار دلالة لعباده على عظمته وارشادا لهم الى حكمته
وحثا لهم على عبادته وحده لا شريك له ليس كمثله شيء وهو خالق
كل شيء وهو العزيز الحكيم ∴ واما بعد فقد سلف من قولنا وما
جمعناه من العلوم في صناعة النجوم في الجزوين الاولين من كتابنا
هذا ما مضى ونحن نذكر في هذا الجزء وما يتبعه ما شاء الله من
مواقيت القرانات الكائنة بين زحل والمشتري وانتقالهما في المثلثات
وما اتى به من هيئة الفلك في الازمان المشهورة منها والاحداث
المذكورة التي حدثت فيها بالنبوات الظاهرة والايات الباهرة
وما حكم به على كل شيء من ذلك وفسره من ادلته • وما ذكره
هرمس وغيره من الاحكام على القرانات وتحاويل سني العالم
مما عالجته الاوائل في احكام سني المواليد وغير هذا مما يتلوه من مدد
الملوك والولاة واوقات الحروب وذكر احوال الاسعار الى
ان نستتمه بمعونة الله وحسن توفيقه حكى ما شاء الله في كتابه
للقرانات ان اول قران وقع بين زحل والمشتري في بدء [illegible]
لخمس مائة وتسع سنين وشهرين واربعة وعشرين يوما مضت من
الف المريخ وانهما واقعا في سبع درجات وست واربعين دقيقة

قال ماشاالله انه لم يجد ولي السنة من المشتري لانه
صاحب الطالع وصاحب مثلثة الشمس وصدرها وهي تدفع قوتها اليه
ومن الزهرة لانها صاحبة بيت القمر وهي في شرفها وهو في
السادس منقول من موضعه وحل في الحادي عشر راجع والمشتري والمريخ
في الحادي عشر ... في العاشر وهما راجعان فذلك رجوع
الثلثة الكواكب في تحويل هذا القران في وسط السماء وبرج روحاني
على هلاك روحاني الارض في القران الثاني من هذا القران لان الطالع

215r

من الحوت وذكر عدد القرانات الواقعة في هذه المثلثة الارضية قرانا قرانا وبعد موضع كل قران على موضع القران الذي قبله درجتين وخمسا وعشرين دقيقة من عشران بين العلة في ذلك ثم القرانات في مثلثة مثلثة يكون مدة كل قران تسع عشرة سنة وعشرة اشهر وواحد عشر يوما • حتى انتهى الى انتقال الممر من الميزان ومثلثة الى العقرب ومثلثة بعد الغير واربع مائة واثنى عشرة سنة وستة اشهر وستة وعشرين يوما وهو الذي يقع فيه الطوفان فكان الاول من قرانات هذا الممر في درجة واربع وعشرين دقيقة من العقرب ومدته تسع عشرة سنة وعشرة اشهر وعشرة ايام وهيئة الفلك عند دخول الشمس الحمل للسنة التي وقع فيها هذا القران الاول من انتقال الممر على ما حكاه ما شاء الله منها على هذه الصورة ☙

سنة ٢٤١٢

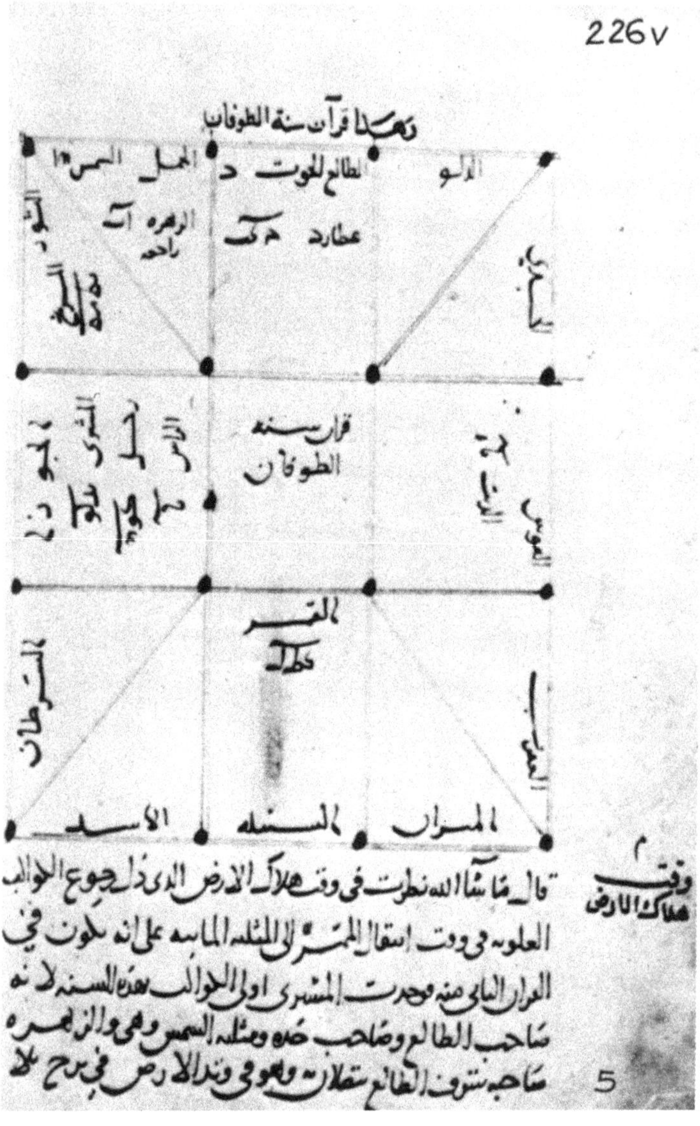
226v
قران سنة الطوفان
الدلو
عطارد
الأسد
السنبله
الميزان
5

226r

برج ذو جسدين وكذلك فانظر في انتقال القمرات من مثلثة الى اخرى
فاذا رايت القمر يؤذن بآفات او هلاك او موالد انها فتامل طالع
سنة الانتقال فان كان ذا جسدين ولا سيما القوس ولكن في
القران الاول شيء ومن القران الثاني بعض فخبر الحادث وان
كان البرج ثابتا او منقلبا فوقت الحادث في القران الاول من
انتقال القمر ولو كان القمر الذي هو صاحب نور الليل في هذه
السنة فابتدا المائي من الدنيا شيء لكنه سليم فدل على النمو والكثرة
بعد الهلاك ووقع القران الثاني من هذا القمر الذي كان ثم منه
الطوفان في ثلث درجات واربعين دقيقة من السرطان وهيئة
الفلك في وقت دخول الشمس الحمل للسنة التي وقع فيها على ما
ثبت في هذه الصورة حسب ما وصفه ما شاء الله من ذلك وحدث
الطوفان في الشهر الخامس منها

227v

من هذه الادله في المتمر يدل على العظايم التي تحدث في القرانات وما يكون منها في وقت القران يدل على ما يحدث في سنته وما يكون منها في وقت تحويل السنه من سني العالم يدل على ما يحدث في شهورها ∴ وكل منزله من هذه المنازل اسهل من التي قبلها فكان بين وقت القران الاول لزحل والمشتري في بدء التحويل وبين شهر الطوفان بعد وضع الشمس سنتين وشهرين واربعه وعشرين يوما الرابع على خمس مايه سنه ذكر ما شا الله انها مضت من الف المريخ اذ كان ما شا الله اجراها في الحساب وسبيلها ان توضع الفان واربع مايه وثلث وعشرون سنه وسته اشهر واما عشر يوما وسوى ما شا الله القرانات بعد ذلك في مثلثه مثلثه الى ان بلغ الى المثلثه الناريه التي كان مولد المسيح عليه السلم في القران التاسع منها وهي مثلثه الاسد فكان القران الثامن من هذا الممر في القوس ثماني عشره درجه وخمسا وخمسين دقيقه ومدته تسع عشره سنه وعشره اشهر وعشره ايام واوله عند تمام خمسه الف وسبع مايه وتسع واربعين سنه وخمسه اشهر وثمنيه ايام من ابتدا القرانات وهيئه الفلك عند دخول الشمس الحمل وفي تحويل سنه العالم التي وقع فيها على ما تثبت في هذه الصوره

227r

بلا بينه مع زحل وان كان له تطلع فيه وصاحب بيته في الطالع
فان ذلك على ما تقدم من صفته من الفساد يكون في هذه السنه
في الشمس ما يسير منها عند بشري بطول المشتري في السرطان
ولولا مكان المشتري وصاحب بيته والقمر الذي له نوبة الليل في الاوتاد
لكان الحادث في السنه الخامسه من الغرق اكثر مما يقع في البلاد
بعرمه وجعله في الشهر الخامس من السنه الاولى منه بقية ما نصير
هذه واعلم انه اذا كانت الكواكب العلويه الثلاثه تحت الشمس في
تحويل انتقالها من مثلثه الى الاخرى لعكس السنين الذي من القرانات
رجعات دلت على انقلابات سمائيه فان كان احد هذه الثلاثه
اسفل من الشمس فيها احسن كان ذلك الانقلابات ارضيه وان
جعت كلها كانت انقلابات سمائيه وارضيه لذلك فانظر فان
كان في بروج بوطانه السنبه وهي الجوزا والسنبله والميزان والنصف
الاول من القوس وجميع الدلو كانت الافات في الناس وان كان في
غيرها من بروج الحيوانات كانت الافات في الاجناس التي تحول
البروج منها . فان اردت ان تعرف صنف الغرق والجدب فيما هو
فانظر عند انتقال القرانات من مثلثه الى الاخرى الى سهم المسر وهو في اي
برج يقع فان وقع في برج ناري كانت الافه من النار وفي برج مائي
كانت الافه من الما وفي برج هوائي يكون العذاب بالريح . وفي
برج ارضي يكون من رجفه او زلازل او سقوط حجاره وما يكون من

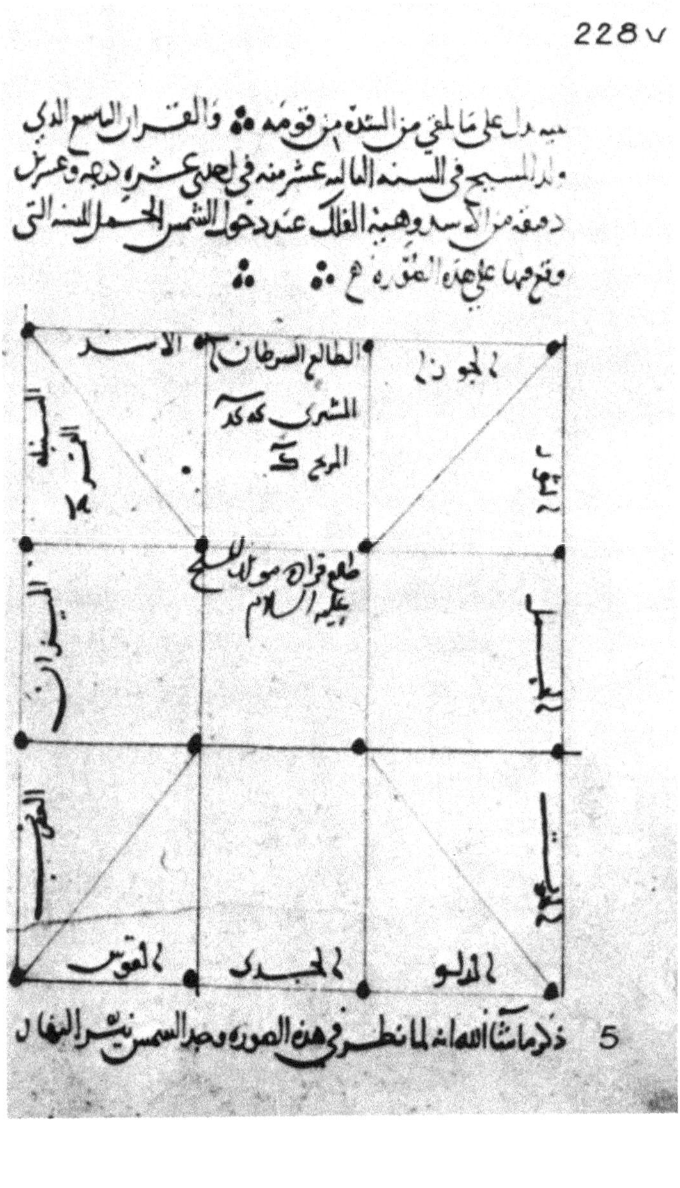
228v
الأسد
الطالع السرطان
المشترى
الجوزا
القوس
الجدى
الدلو
5

Text

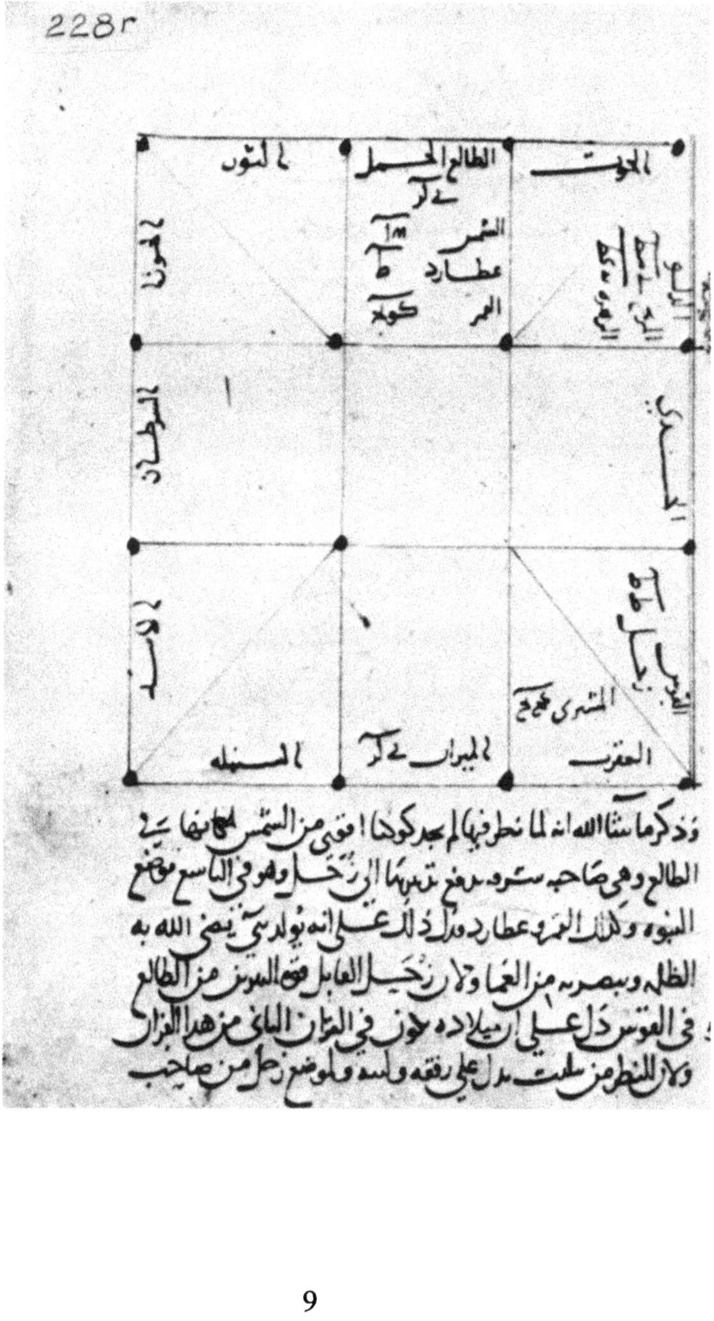
228r
الطالع الحمل
الحوت
الثور
الشمس
عطارد
القمر
الميزان
العقرب
السنبله
المشتري
وذكر ماشاالله انه لما نظر فيها لم يجد كوكبا اقوى من الشمس لكونها في
الطالع وهي صاحبة شرفه فرفع تدبيرها الى زحل وهو في التاسع موضع
النبوة وكذلك القمر وعطارد فدل ذلك على انه يولد نبي يمضي الله به
الظلمة ويبصر به من العمى ولان زحل القابل فوق الشمس من الطالع
في القوس دل على ان ميلاده يكون في القران الثاني من هذا القران
ولان النظر من تثليث يدل على رفقه ولينه ولموضع زحل من صاحب

229v

ثم انحدر في مواضع كونهما وجد صاحب السنة زحل العقوة وكثرة
شهادته ولان النحوس فيها سوى المريخ والقمر كانت تدفع بينهما اليه
وهو في بيته والطالع شرفه ومثلثه وصاحب الطالع يدفع تدبيره
اليه من بيته الثاني دل ذلك على انه يولد في عشرة اشهر من هذه السنة
ولان النحوس السفلي والقمر مع المريخ في التاسع دل على شيء يلقاه وخاف
عليه القتل في مكان من يدفع لانهما مقترنان في عاشر البروج فيصير
من وقت القران الاول في بدء التحويل الى السنة التي ولد فيها المسيح عليه
السلام وهي الثالثة عشر من هذا القران بعد اسقاط الستة سنين والشهور الرابعة
على خمسمائة سنة الماضية من الف المريخ اذ كان ماشاء الله اجراها
في الحساب ليس ... فيه خمسة الف وسبعمائة وثلث
وسبعون سنة واربعة وعشرون يوما ومن وقت الطوفان والحشر
السنة التي ولد فيها المسيح ثلثة الف وثلثمائة وتسع واربعون سنة
خمسية وستة اشهر واحد عشر يوما ومضت القرانات بعد ذلك
على توالیها حتى انتقل القمر من المثلثة الهوائية التي هي الجوزاء ودوا ...
الى السرطان ومثلثه التي كانت فيها دولة العرب عند تمام ستة الف
وثلثمائة وخمس واربعين سنة وثلثة اشهر وعشرين يوما ووضعنا فيها
الستة سنين والشهور الرابعة على الخمسمائة سنة الماضية من الف المريخ
فهي ستة الف وثلثمائة وستة وثلثون سنة وستة وعشرون يوما من
وقت القران الواقع في بدء التحويل وكان الاول من قرانات هذه

229r

وعطارد صاحب بيت العمر يقارن تدبيرهما الى زحل صاحب
السابع والثامن من الطالع وهو مقبول في موضعه مائل الاستقامه
فدل ذلك على ان النبي الذي دل القران الاول على ميلاده في هذا القران
يولد عند تمام ثلث عشرة سنة منه وانه وجد هيئة الفلك
في وقت تحويل سنة العالم التي ولد فيها المسيح عليه السلم وهي الثالثة
عشر على هذه الصفة

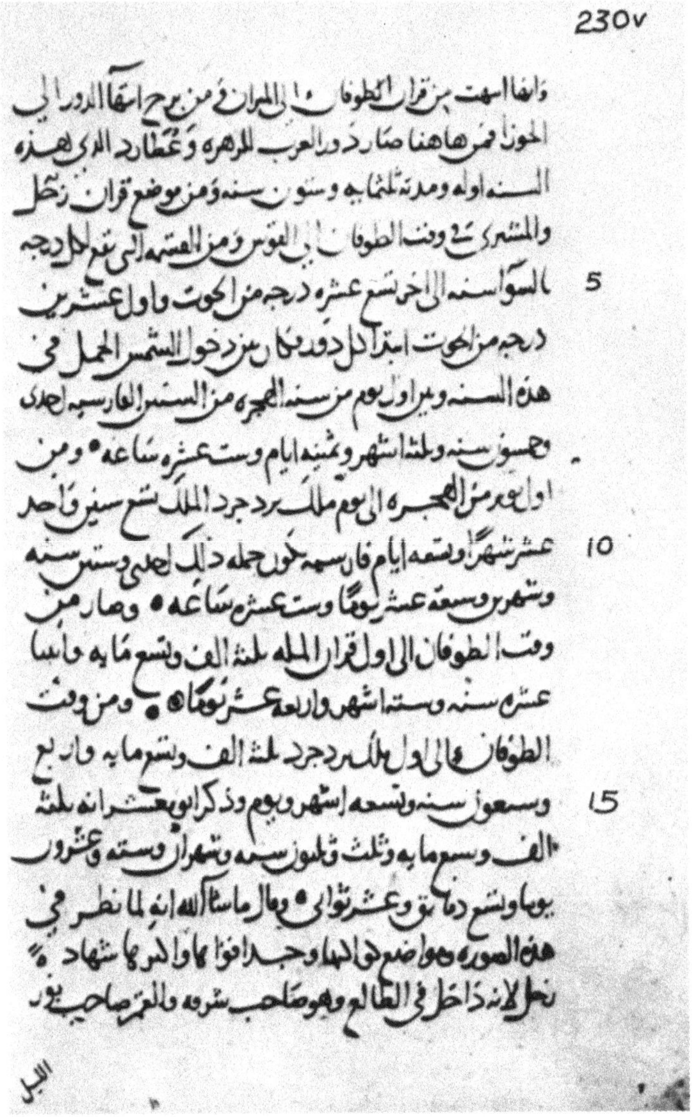

230v

وانما اسقطت من قران الطوفان الى الميزان في من برج اسقاط الدور الى
الحوت ثم من هاهنا صار دور العرب للزهرة وعطارد الذي هذه
السنة اوله ومدته ثلثماية وستون سنة ومن موضع قران زحل
والمشتري في وقت الطوفان في القوس ومن القسمة الى تتم لكل درجة
السنوات سنة الى اخر تسع عشرة درجة من الحوت واول عشرين
درجة من الحوت ابتدا دل دور وكان من دخول الشمس الحمل في
هذه السنة ومن اول يوم من سنة الهجرة من السنين الفارسية احدى
وخمسون سنة وثلثة اشهر وثمنية ايام وست عشرة ساعة ومن
اول عمر من الهجرة الى يوم ملك يزدجرد الملك تسع سنين واحد
عشر شهرا وتسعة ايام فارسية تكون جملة ذلك احدى وستين سنة
وشهرين وتسعة عشر يوما وست عشرة ساعة وصار من
وقت الطوفان الى اول قران الملة ثلثة الف وتسع ماية وخمسا
عشرة سنة وستة اشهر واربعة عشر يوما ومن وقت
الطوفان الى اول ملك يزدجرد ثلثة الف وتسع ماية واربع
وسبعون سنة وتسعة اشهر ويوم وذكر ابو معشر انه ثلثة
الف وسبع ماية وثلث وثلثون سنة وشهران وستة وعشرون
يوما وتسع دقائق وعشر ثواني وقال ماشاالله انه لما نظر في
هذه الصورة بمواضع الى كانها وجد اقوى ما واكثرها شهادة
زحل لانه ناظر في الطالع وهو صاحب شرفه والقمر صاحب بيت

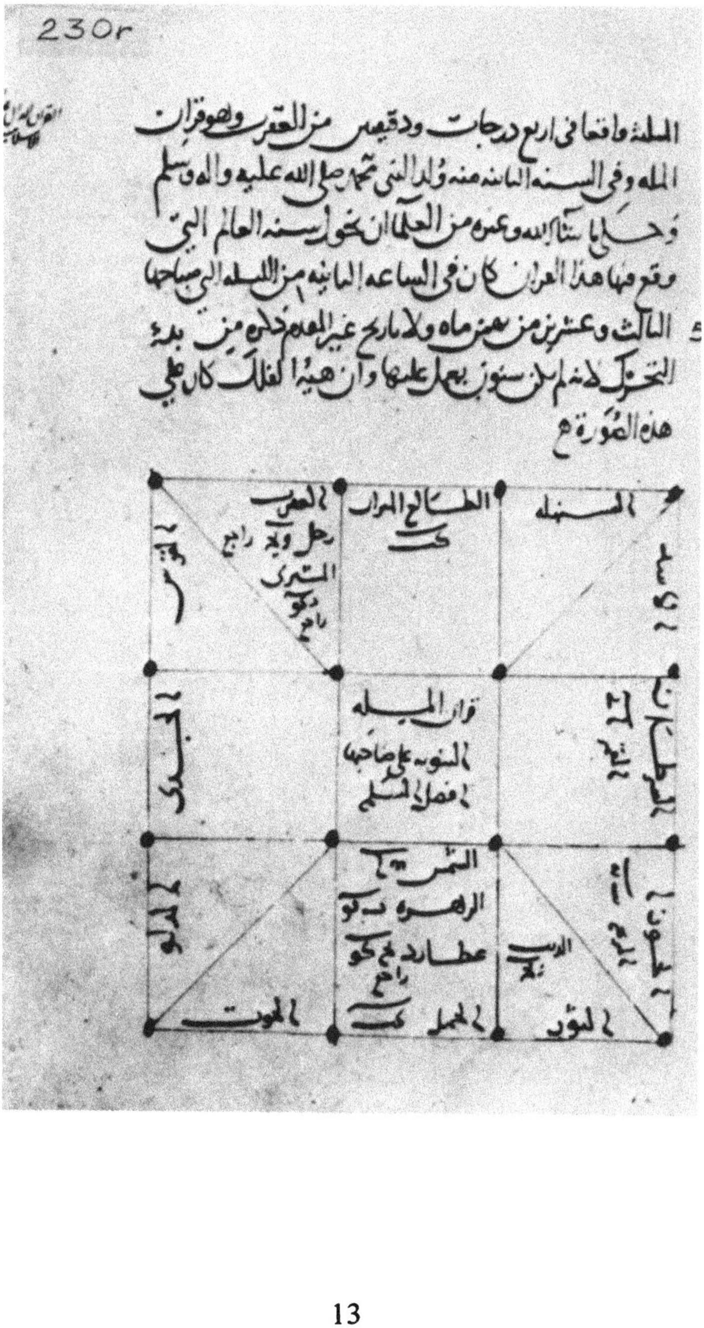
230r

المثلثة واقعا في اربع درجات ودقيقتين من العقرب وهو قران
الملة وفي السنة الثانية منه وُلد النبي محمد صلى الله عليه واله وسلم
وحكى ما شاء الله وغيره من العلما ان تحويل سنة العالم التي
وقع فيها هذا القران كان في الساعة الثانية من الليلة التي صباحها
الثالث وعشرين من [illegible] ماه ولا تاريخ غير العلم ذلك من بدء
التحرك لانه لم يكن يشعرون بعمل عليها وان هيئة الفلك كان على
هذه الصورة

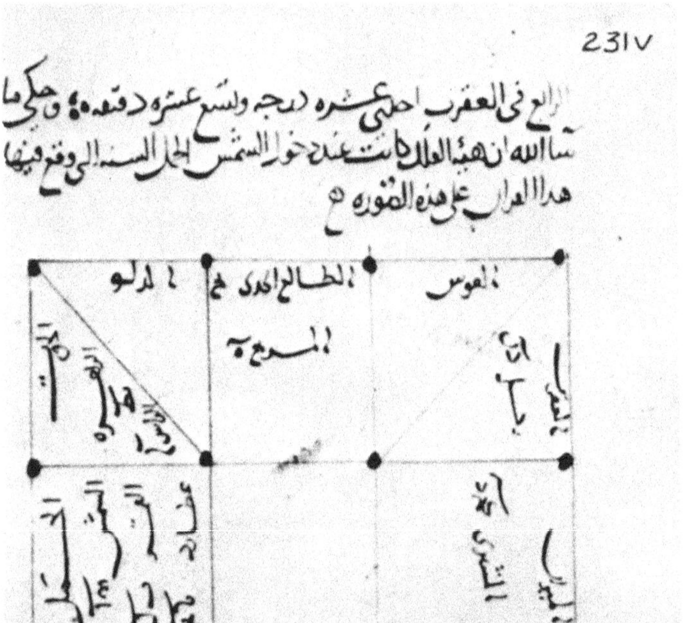
231V

الرابع في العقرب احدى عشره درجه وتسع عشره دقيقه، وحكى ما
شا الله ان هيئة الفلك كانت عند دخول الشمس الحمل السنه التي وقع فيها
هذا القران على هذه الصوره

فانه نظر الى طالعها ومواضع كواكبها حيث رجع القران الى برج الاصل
الذي دل على ميلاد النبي صلى الله عليه واله وسلم فدل موضع التحسين
في ودرس ورجال الشمس والقمر والمشتري على الاوتاد وقوه زحل

231r

الليل يدفع تدبيره اليه من العاشر بالعدد والتاسع بالقسمه مكان البنوه
والمشترى يدفع تدبيره اليه ⁂ ووجد المريخ في الجوزا وهو التاسع بالعدد
في اعلى المنطقة يصعد الى غايتها مقبل يدبر الشمس والزهره وعطارد
رجعته مثله لك اجمع ومكان السهم الاول وهو خال في العقرب وصاحبه
في بيت الدين على انه يولد في السنه الثانيه من هذا القران نبي في تهامه
ارض العرب وذلك لمكان السهم الاول في العقرب واتصال القمر بالزهره ولو
كان الطالع برجا ذا جسدين وحل برج ذي جسدين لقلت ان ميلاده
يكون في القران الثاني لقبول الزهره والمريخ في مواضعهما ومكان القمر في
وسط السما يدل على انه يسلم من القتل ولان الزهره قد جاوزت الاحتراق
تريد الخروج من الشعاع يدل على انه يلقى شدائد ويتولى حينا ثم يعلو
ويقوى ويذكر وملكه اهل ملته لمكان القمر في وسط السما والمريخ في بيت
الدين يدل على انه يطلب الدين واقامته بالقتال ولاتصال القمر بالزهره
يكون في اهل ملته طبايع الزهره واخلاقها ولمعنى اهل تهامه في السنه
الثانيه من هذا القران شيه من العدد حتى يهربوا ويتفرقوا في البلاد وكذا
كان في هذه السنه عام الفيل وهي التي ولد فيها النبي صلى الله عليه وسلم
ثم مضت سنو القران الاول وهي تسع عشره سنه وعشره اشهر وعشره ايام
وسنو القران الثاني وسنو القران الثالث فصار جميع ذلك تسعا وخمسين
سنه وسبعه اشهر اخرها قبل الوقت الذي هلك فيه يزدجرد لسنه
وتسعه اشهر وسبعه عشر يوما وست عشره ساعه ⁂ ووقع القران
١٢ ٩

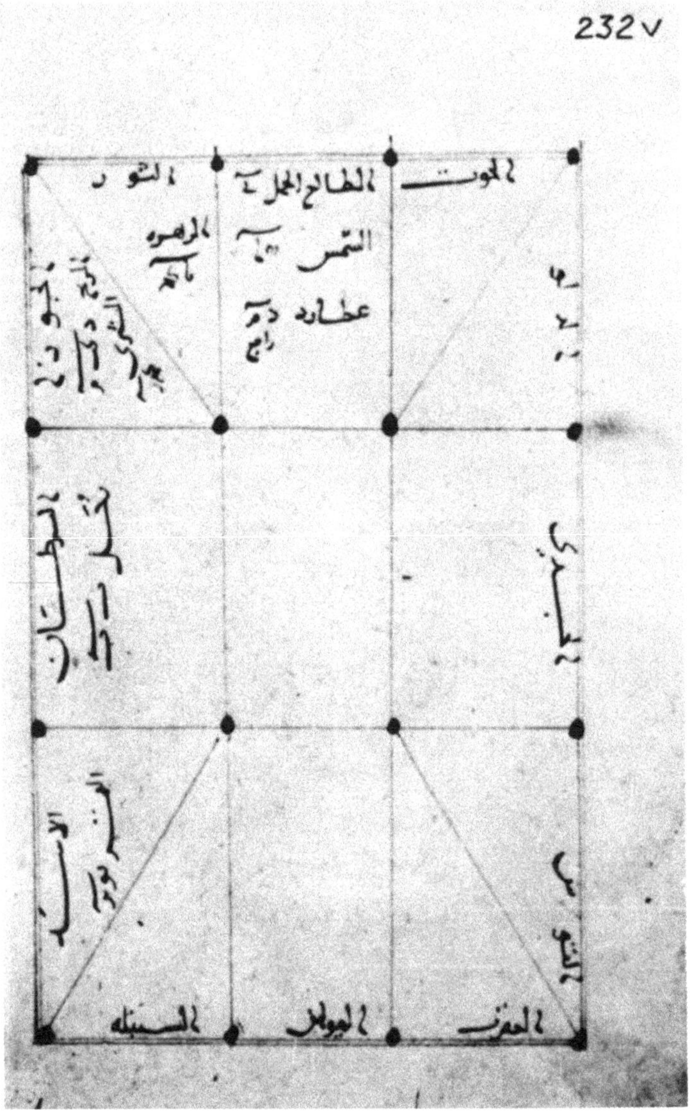
232v
الحوت
الطالع الحمل
الثور
الزهرة
الشمس
عطارد
العقرب
الميزان
السنبلة

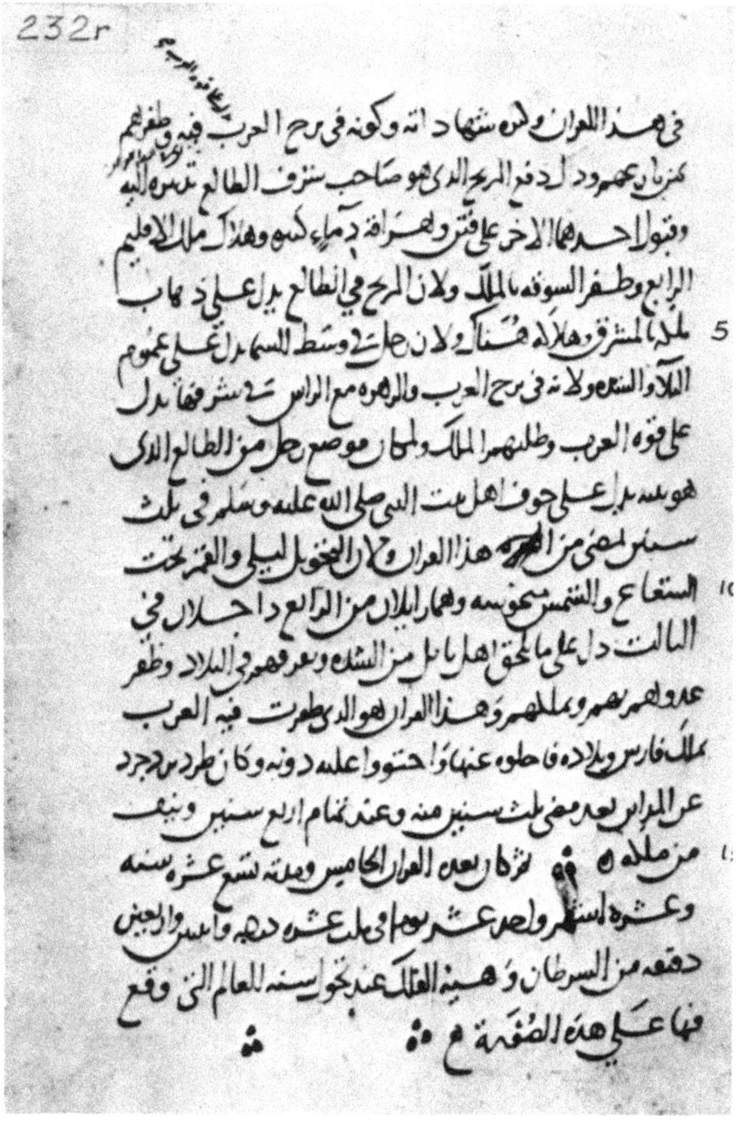

232r

في هذا القران ولشهادته وكونه في برج العقرب فيه ظفرهم
بمن يعاديهم لدفع المريخ الذي هو صاحب شرف الطالع بتسييره اليه
وقبول احدهما الاخر على قتل وهراقة دماء كثيرة وهلاك ملك الاقليم
الرابع وظفر السوقة بالملك ولان المريخ في الطالع يدل على ذهاب
ملك المشرق وهلاكه هناك ولان حل في وسط السماء دل على عموم
البلاء والشر ولانه في برج العقرب والزهرة مع الراس في شرفها يدل
على قوة العرب وطلبهم الملك ولمكان موضع حل من الطالع الذي
هو بيته يدل على خوف اهل بيت النبي صلى الله عليه وسلم في ثلث
سنين لمضي من الهجرة هذا القران ولان العقول للبيلي والقمر تحت
الشعاع والشمس منحوسة وهما ايلان من الرابع داخلان في
الثالث دل على ما لحق اهل بابل من الشك وتفرقهم في البلاد وظفر
عدوهم بهم وبملكهم وهذا القران هو الذي طردت فيه العرب
ملك فارس وبلاده فاجلوه عنها واحتووا عليه دونه وكان طرد يزدجرد
عن المداين بعد مضي ثلث سنين منه وعند تمام اربع سنين وشيء
من ملكه ۞ ثم كان بعد القران الخامس ومائة تسع عشرة سنة
وعشرة اشهر واحد عشر يوما في ثلث عشرة درجة واربعين والعشرين
دقيقة من السرطان وهيئة الفلك عند تحويل سنة العالم التي وقع
فيها على هذه الصفة ۞

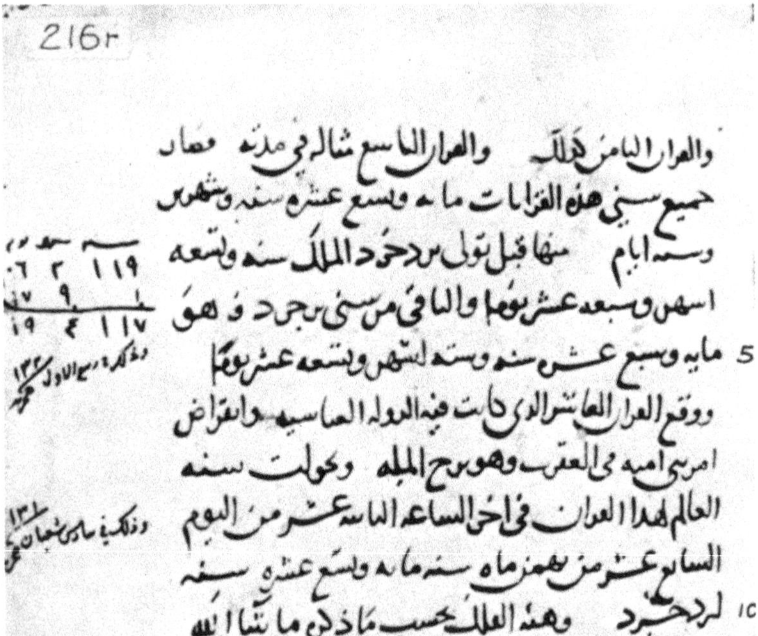

216r

والقران الثامن هلك والقران التاسع شمالي في مدينة فصار
جميع سني هذه القرانات مائة وتسع عشرة سنة وشهر
وستة ايام منها قبل تولي يزدجرد الملك سنة وتسعة
اشهر وسبعة عشر يوما والباقي من سني يزدجرد وهو
مائة وسبع عشرة سنة وستة اشهر وتسعة عشر يوما
ووقع القران العاشر الذي كانت فيه الدولة العباسية وانقراض
امر بني امية في العقرب وهو برج الملة وتحولت سنة
العالم لهذا القران في اخر الساعة التاسعة عشر من اليوم
السابع عشر من بهمن ماه سنة مائة وتسع عشرة سنة
ليزدجرد وهذه العلل بحسب ما ذكر ما شاء الله
منها على هذه الصورة

233v

. جبر ما شا الله انه نظر في هذا التحويل الى الطالع ومواضع الكواكب وكان صاحب السنة الشمس وهي تدفع تدبيرها الى زحل من تربيع فدل ذلك على هلاك ملك بابل واستيصاله في السنة الرابعة منه وحدوث فتن في اخر السنة التاسعة وفساد وهراقة دما كثيرة لان زحل هو الى تدبير الشمس من تربيع لانها في شرفه ومانعها الملك وعم دفع الشمس وهي صاحبة السنة تدبيرها الى زحل ما دلت على استيصال ملك فارس على قتال وحروب ولكنها دفعت اليه التدبير فلما صار في شرفه مقابلا لانها نازعها الملك فدلت على ما ذكرت ولم يزل كذلك حتى يخرج من شرفه ويصير الى العقرب ويموت صاحبه المشرق الصابر وعند التروع في السنة الاولى والثانية • ولما كان القران الاعظم منقلبا من الحوت الى السرطان ومن الميزان الى العقرب ومن الدلو الى الحوت دلت على انه يغني من يشو هذا الملك الهالك ومن اهله من كان في الارض التي تتولاها الحوت بطن سنان وما يليها ولسقوط المشتري عن الوتد مكانه في الجوزا مع المريخ وفساد الشمس بزحل والقمر بالذنب دل على هلاك ملوك الناس وعموم هلاكهم وهراقة دما كثيرة في ارض السرطان التي هي الاقليم الرابع ❀

ومن بعد هذا القران السادس ومدته تسع عشرة سنة وعشرة اشهر واحد عشر يوما ❀ والقران السابع ومدته كذلك

والقران

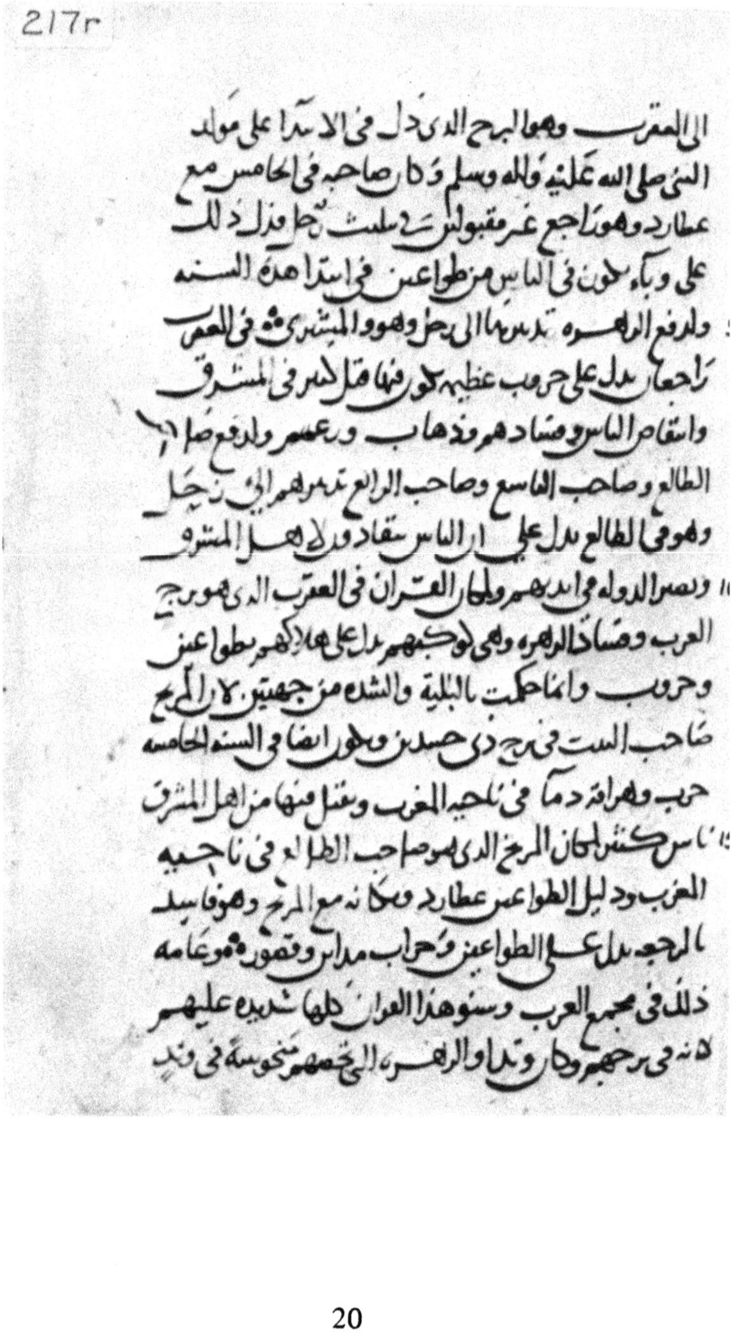
217r

الى العقرب وهو البرج الذي دل في الابتدا على مولد
النبي صلى الله عليه وآله وسلم وكان صاحبه في الخامس مع
عطارد وهو راجع غير مقبول بسبب تثليث زحل فدل ذلك
على وباء يكون في الناس من طواعين في ابتدا هذه السنه
ولدفع الزهره تدبيرها الى زحل وهو والمشتري في العقرب
راجعان يدل على حروب عظيمه يكون فيها قتل كثير في المشرق
وانتقاص الناس وفسادهم وذهاب وعسر ولدفع صاحب
الطالع وصاحب التاسع وصاحب الرابع تدبيرهم الى زحل
وهو في الطالع يدل على ان الناس ينقادون لاهل المشرق
وتنصر الدوله في ايديهم ولمكان القران في العقرب الذي هو برج
العرب وفساد الزهره وهي كوكبهم يدل على هلاكهم بطواعين
وحروب وانما حكمت بالبلية والشدة من جهتين لان المريخ
صاحب البيت في برج ذي جسدين ويكون ايضا في السنه الخامسه
حرب وهراقة دما في ناحيه المغرب ويقتل فيها من اهل المشرق
ناس كثير لمكان المريخ الذي هو صاحب الطالع في ناحيه
المغرب ودليل الطواعين عطارد ومكانه مع المريخ وهو فاسد
الرجعة يدل على الطواعين وخراب مدائن وقصور وهو عامه
ذلك في مجمع العرب وسنو هذا القران كلها شديده عليهم
لانه في برجهم وكان وتدا والزهره التي تخصهم منحوسة في وتد

216 v

الطالع العقرب
زحل
المشتري
الميزان
القوس
قران الاولة العباسية
الزهرة
الشمس
عطارد
المريخ
الحمل
الثور
الجوزا

واسقطت السنة الرابعة من قران الملة الى السرطان ومن موضع
القران الى الاسد د من درجة العشية الواقعة في اول دلو وقد
في عشرين درجة من الحوت الى تسع عشرة من السنبلة
وفيه شعاع المريخ وبمطارحه بالرجعة فذكر ما شا
الله انه ينظر في تحويل هذه السنة التي رجع فيها القران

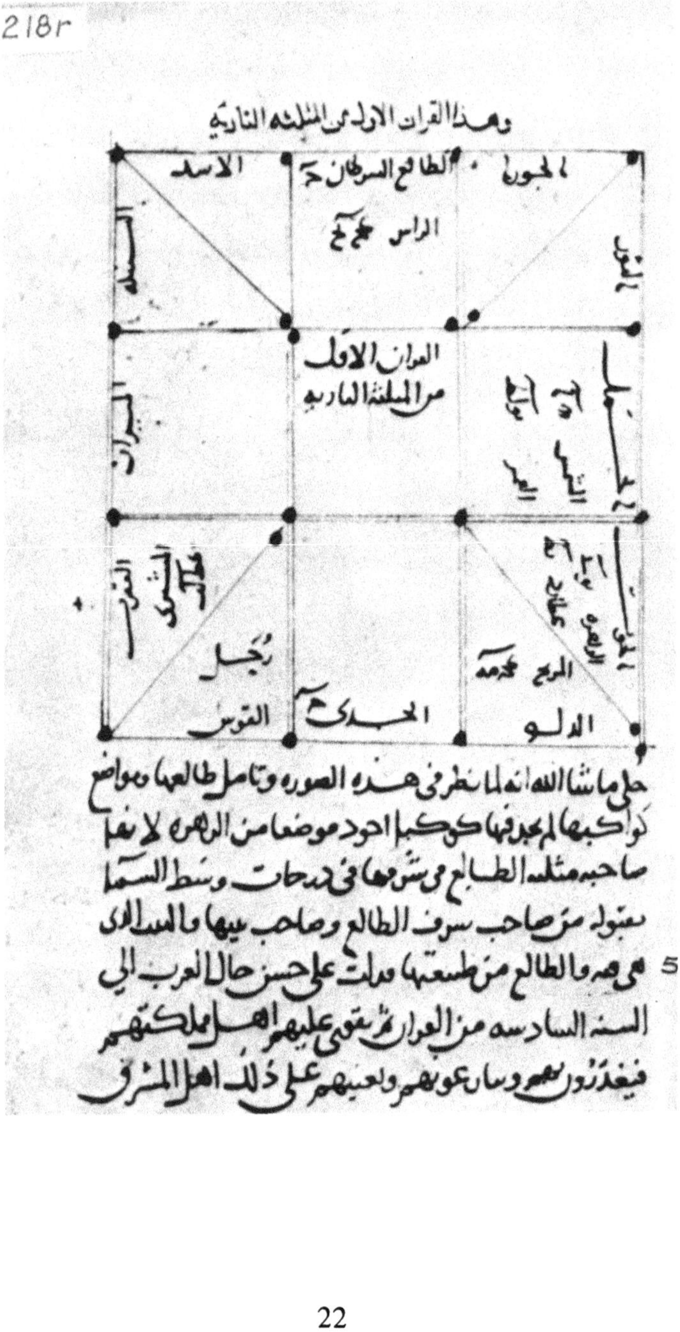
218r

وهذا القران الاول من المثلثه الناريه

حلى ماشاالله انه لما نظر في هذه الصوره وتامل طالعها ومواضع
كواكبها لم يجد فيها كوكبا احود موضعا من الزهره لانها
صاحبه مثلثه الطالع وهي في شرفها في درجات وسط السما
مقبوله من صاحب شرف الطالع وصاحب بيتها والمشترى
في صدر الطالع من طبيعتها دلت على حسن حال العرب الى
السنه السادسه من القران ثم يقوى عليهم اهل مملكتهم
فيغدرون بهم ويسانعونهم ويعيبهم على ذلك اهل المشرق

217v

احمر مقابله زحل وهو راجع . وصاحب وسط سمايهم
عطارد وكان راجعا منحوسا فدل على الشده من وجوه
كثيره من الموت والقتل ومن جور السلطان وعسفه .
ثم كان بعد هذا القران العاشر الذي انتقلت فيه الدوله والمملكه
عن بني اميه الى بني العباس وحدثت فيه الحروب وسفكت
فيه الدماء وخربت المدن القران الحادي عشر . والقران
الثاني عشر وهو اخر قرانات مثلثة العقرب تكون جميع سني هذه
الثلثة القرانات تسعا وخمسين سنة وسبعة اشهر وثلثة ايام
اخرها عند تمام مايه وسبع وستين سنه وشهر وعشرين
يوما ليزدجرد وانتقلت القرانات الى مثلثة القوس النارية في هذا
الوقت وكان تحويل سنة العالم التي وقع فيها القران الاول
منها على سبع ساعات ونصف من اليوم الثاني وعشرين من
بهمنماه سنه مايه وسبع وستين ليزدجرد وهيلاج الفلك
على ما ثبت في هذه الصوره وانتهت هذه السنه من قران الطالع
المليه الى السرطان ومن موضع القران فيه الى الاسد . ومن
القسمه الى تسع عشره درجه من العقرب .

219r

خراسان والترك وما والى تلك البلاد على العرب من غير
ان يخرج الملك عنهم للوصلة التي قدمنا ذكرها من القرانين
ويشرحنا حالها وذلك كذلك الى ان ينتقل القران من مثلثة
القوس الى مثلثة اخرى لا يكون من طالعها وبين اوتاد هذه
المثلثة مناسبة ولا وصلة ∴ ويستمر انتقال القران زحل
والمشترى فيها بحسب ما جرى في ابتداء القوس عند انتقال القرانات
من مثلثة الحوت الى مثلثة العقرب ولم يكن بين المثلثتين مناسبة
ولا مناسبة دل على ذهاب دولة فارس وظهور دولة العرب
وانتقال الملك اليهم ولم يزل الامر على الاضطراب والتحيف
الى ان استمرت القرانات في المثلثة وعادت الى العقرب في
القران الرابع فذهب ملك فارس واستعلى ملك العرب حيث
لم يكن بين المثلثتين مناسبة ولا مواصلة فاما هذا الانتقال المواصل
كما وصفنا الذي قبله فان ماشاء الله حكى ان اعمار ملوكه تطول
في مثلثة العقرب لقوة دلالة سلطان القوس على الملك وتضطرب
المملكة والدولة في اخر سني القران السابع حين يقترن المريخ وزحل
في السرطان ويرجع التدبير الى القوس فعند ذلك خاف ترك الملك
الذي يأتي في هذا الزمان خمس عشرة سنة ويخرج السيف في اخر
امره ويملك في الناس ∴ وفي السنة التاسعة عشر من القران
السابع يغزو الناس بعضهم بعضا وتكثر الحروب الى القران الثامن

واسهان لمجاز القران في القوس وينتقل الملك من بيت الى بيت في السنه الرابعه من القران وكثر الموت في السنه العاشره منه وينتقل الدوله الى اهل المشرق لانتقال القران الى القوس وتطول اعمار ملوك هذه المملكه ويجرد السيف في الادميين لان المريخ وهو صاحب الحادي عشر كان في الثامن في برج السنه وبعض الكواكب منه ولولا القبول لكان الامر عظيما شديدا ولكن القبول خفف الشر ولان طالع هذا القران في اول الانتقال وقع في السرطان وهو وتد من اوتا د طالع القمر الاول الذي كان في العقرب وتثليثها ودل على مولد النبي صلى الله عليه وسلم وعلى دوله العرب وانتقال الملك اليهم وبيت السلطان منه وبيت القمر وكان القمر في هذا القران في وسط السما ومتمكنا من الوتد مستوليا في مكانه من صاحب بيته زايد النور فدل على ثبات هذه الدوله في يد اهلها وبقا سلطانهم فيهم لان السرطان طالع هذا الانتقال هو بيت السلطان من القران الماضي وثبات الدوله في ايدي العرب منه والانتقال من شكل الى اخرى مما انتقلت من بطن الى بطن ومن قوم الى اخرين ولانه في برج مشرقي دلت الدوله للملك المشرقي والطمانينه والانتقال اليه ولان القرانات تدور في القوس والاسد والحمل دلت الغلبه للمشارعه من اهل

220r

الطالع الاسد في اخر درجة منه والمريخ معه في البرج
الملس عن الوتد والقمر في الطالع يدفع الى عطارد والشمس
في التاسع تدفع الى زحل والمشتري في سنبله ساقطا عن
البيت الحادي عشر وعطارد في الثامن داخلا في الظهيرة ⁂
والزهرة في وسط السماء تدفع تدبيرها وقوتها الى المريخ الذي
هو ضدهما في الفلك والبيت وهي غير مقبولة وكان عطارد في
دره وفي هبوطه فدل ما وصفنا على قتل وموت كثير ناحية
المشرق ويظهر في اول سني القران خارج للسلطان في هذه النواحي
ويدخل عليه ويحارب السلطان الاول ويدعو الناس الى دين غير
دينهم ويقتلون ويذهب امانهم ويطلبون الباطل وتكلب
الرعية ويذهب هيبة السلطان والملوك من قلوبهم ويقوى
الاشرار ويضعف الاخيار ويموت اكثرهم ويستعلي النسا
على الرجال ويفشو الفاحشة وترتفع الرحمة ويأتي الشر من
قبل المشرق وبه يقوى ويكون ظهوره من ناحيته في اول
هذه السنة ويدعي علم النجوم كذبا وعاقبة امره الى فساد
وجميع ذلك في السنة الاولى من القران ⁂ وفي الثانية
يكون قتال وسفك دما ⁂ وفي الثالثة موت وفنا بالوبا
وفساد الزروع ⁂ وفي السنة الثامنة يكون ايضا قتل ⁂
وفي التاسعة يكثر الجراد ويفسد الزروع والنبات وكون الاقاليم

219v
5

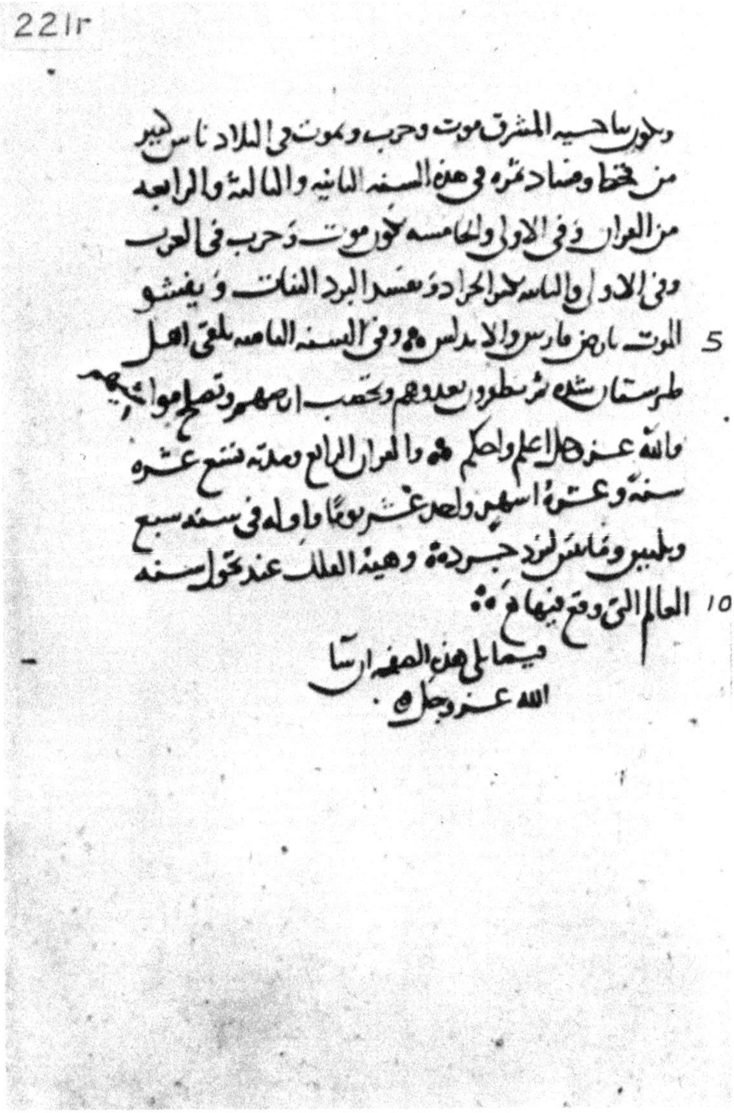
221r

ويكون بناحيه المشرق موت وحرب وموت في البلاد ناس كثير
من قحط وفساد ثمره في هذه السنه الثانيه والثالثه والرابعه
من القران وفي الاولى والخامسه يكون موت وحرب في العرب
وفي الاولى والسادسه يكثر الجراد ويعسر البرد النبات ويفشو
الموت بارض فارس والاندلس ٭ وفي السنه التاسعه يلقى اهل
طبرستان شده ثم يظفرون بعدوهم ويخصب ارضهم وتصلح مواشيهم
والله عز وجل اعلم واحكم ٭ والقران الرابع ومدته تسع عشره
سنه وعشره اشهر واحد عشر يوما واوله في سنه سبع
وثلثين ومائتين ليزدجرد وهيئة الفلك عند تحويل سنه
العالم التي وقع فيها ٭
فيما يلي هذه الصفحه ان شا
الله عز وجل

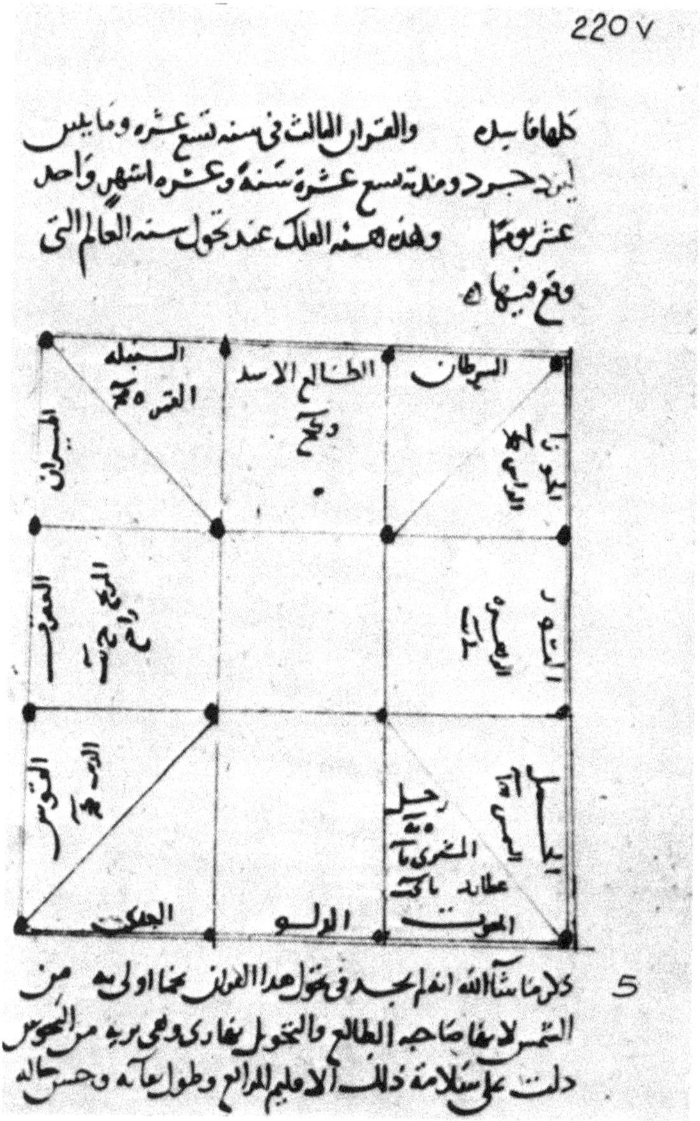

220 v

كلها فاسد والقران الثالث فى سنه تسع عشره وما يليس
ليزدجرد ومدته تسع عشرة سنة وعشره اشهر واحد
عشر يوما وهذه هيئه الفلك عند تحول سنه العالم التى
وقع فيها ه

5 ذكر ما شا الله انه لم يجد فى تحول هذا القران نجما اولى به من
الشمس لانها صاحبه الطالع والتحويل نهارى وهى بريه من النحوس
دلت على سلامة ذلك الاقليم الرابع وطول عمره وحسن حاله

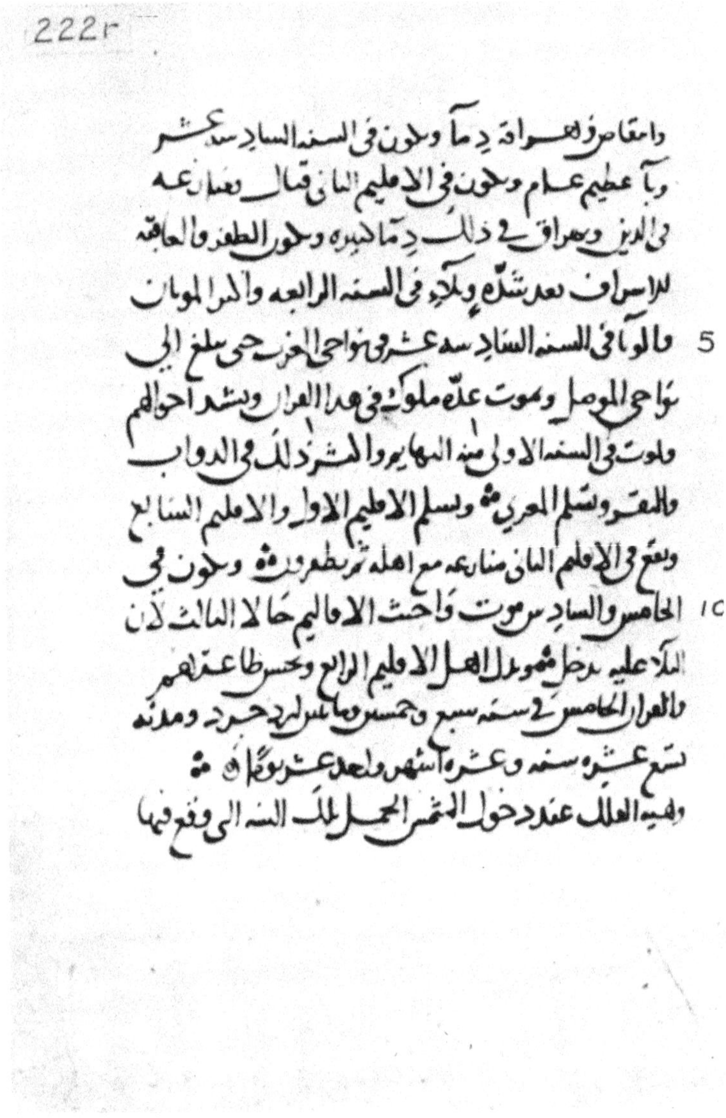

222r

وانتقاص في هواه ودماء ويكون في السنة السادسة عشر
وباء عظيم عام ويكون في الاقليم الثاني قتال يعظم غمه
في الدين ويهراق في ذلك دماء كثيرة ويكون الطعن والعاهة
للاشراف بعد شدة وبلاء في السنة الرابعة واكثر الموتان
والوباء في السنة السادسة عشر في نواحي العرب حتى يبلغ الى
نواحي الموصل ويموت عدة ملوك في هذا القران ويشتد احوالهم
ويموت في السنة الاولى منه البهائم واكثر ذلك في الدواب
والبقر ويسلم العرب ويسلم الاقليم الاول والاقليم السابع
ويقع في الاقليم الثاني منازعة مع اهله ثم يظفرون ويكون في
الخامس والسادس موت واحسن الاقاليم حالا الثالث لان
الملك عليه يدخل سهم دل اهل الاقليم الرابع ويسقط عنهم
والقران الخامس في سنة سبع وخمسين ومائتين ليزدجرد ومدته
تسع عشرة سنة وعشرة اشهر واحد عشر يوما
وهيئة الفلك عند دخول الشمس الحمل لملك السنة التي وقع فيها

221v

ذكر ماشاالله ان القران يجي في القوس واوتاد الطالع ثابته والجنوبي مر
راليه الا الشمس فانها في الثامن وبينها وبين مركز السابع
درجات فيقع دبيرها الى زحل ثه والمريخ في التاسع ثه والقمر في
الخامس فيقع تدبيره الى المشتري عند تحويل فدل ذلك على ان
ملك الاقليم الرابع يهلك في السنة الرابعة من هذا القران
ويقتل الملك من اهل بيته ويكون باصبهان ونواحيها ملحمة شديده
وضياع ملك الاقليم الرابع ويقع منازعة في ارض العرب

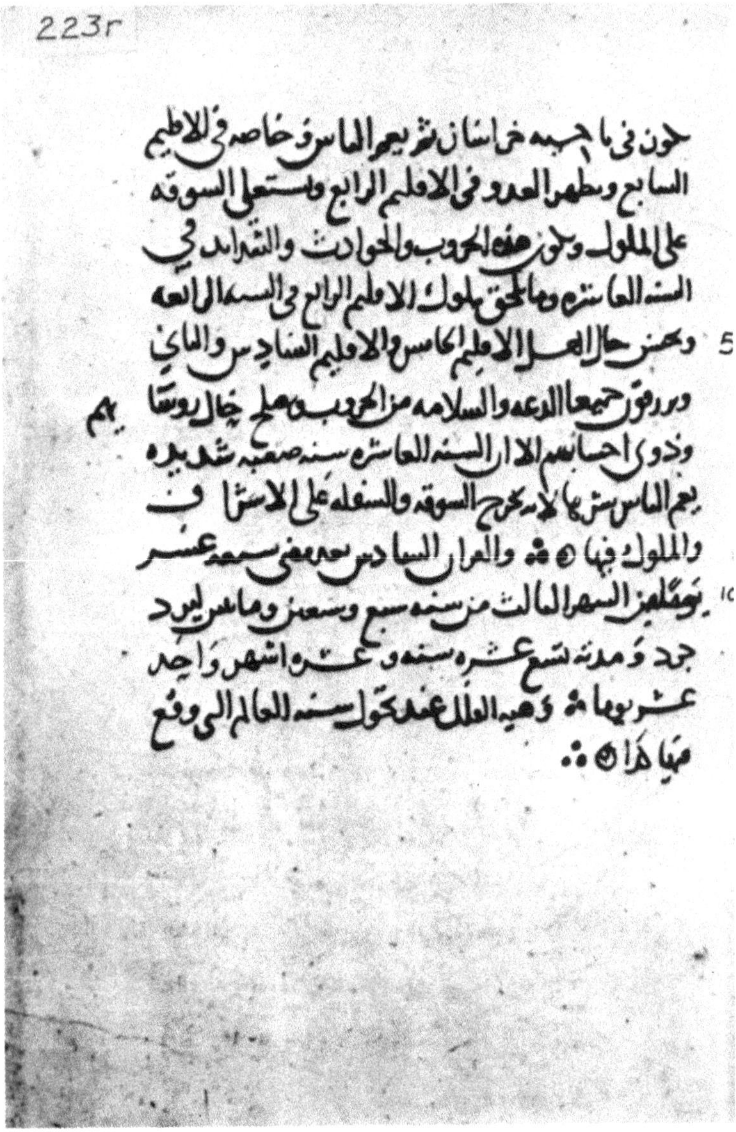

223r

يكون في ما جهة خراسان ثم يعم الناس خاصة في الاقليم
السابع ويظهر العدو في الاقليم الرابع ويستعلي السوقة
على الملوك ويكون عند الحروب والحوادث والشدائد في
السنة العاشرة بها لحوق ملوك الاقليم الرابع في السنة الثالثة
ويحسن حال اهل الاقليم الخامس والاقليم السادس والثاني
ويرزقون جميعا الدعة والسلامة من الحروب الا خال رؤسا يهم
وذوي احسابهم الا ان السنة العاشرة سنة صعبة شديدة
يعم الناس شرها لانه يخرج السوقة والسفلة على الاشراف
والملوك فيها ۞ والقران السادس بعد مضي سبعة عشر
يوما من الشهر الثالث من سنة سبع وسبعين ومائتين ليزد
جرد ومدته تسع عشرة سنة وعشرة اشهر واحد
عشر يوما ۞ وهيئة الفلك عند تحويل سنة العالم التي وقع
فيها هذا ۞

222v

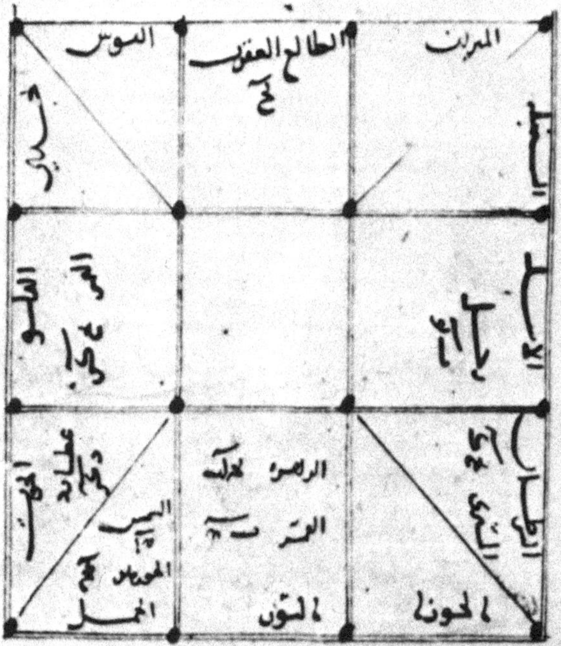

ذكر ما شا الله انه لما نظر في تحويل سنة هذا القران وجد
المريخ اشد الكواكب سهاما فيه والزهرة اقوى موضعا وهي
مقبلة في مكانها الى المشتري وهو غير مقبول والقمر نور الليل
مقبلة بقوة الى زحل مستقبلها منه يدل مكان الشمس والقمر
من زحل على شدائد تصيب الملوك في هذا القران من
امراض وحروب ولهم اعداء وعلى حرب عظيمه وقتل

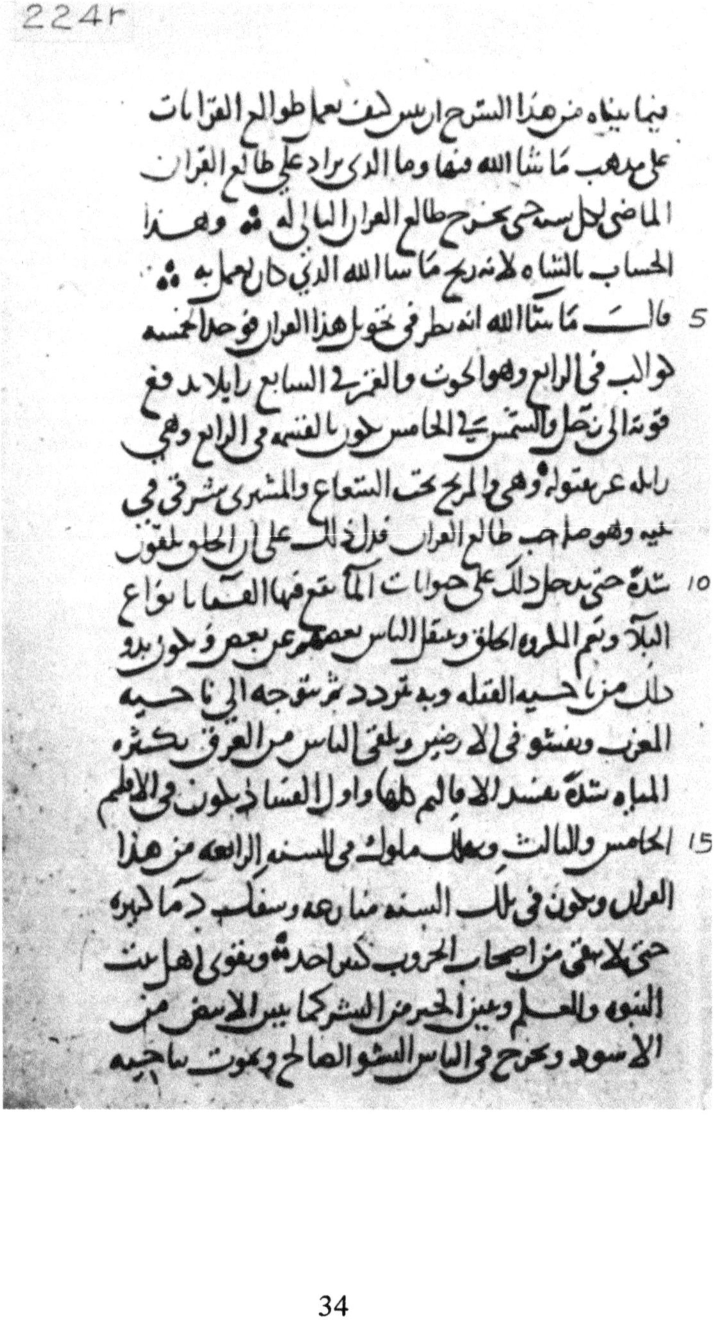

224r

فيما بيناه من هذا الشرح ارس كيف نعمل طوالع القرانات
على مذهب ما شا الله فيها وما الذي يراد على طالع القران
الماضي لكل سنة حتى يخرج طالع القران الثاني له ∴ وهذا
الحساب بالشاه لانه ريح ما شا الله الذي كان يعمل به ∴
قال ما شا الله انه نظر في تحويل هذا القران فوجد الخمسة
كواكب في الرابع وهو الحوت والقمر في السابع زايلا وقد دفع
قوته الى زحل والشمس في الخامس تكون بالقسمة في الرابع وهي
زايلة عن قوتها وهي والمريخ تحت الشعاع والمشتري شرقي في
حيزه وهو صاحب طالع القران فدل ذلك على ان الخلق يلقون
شدة حتى يدخل ذلك على حيوانات الماء يعرفها القسما بانواع
البلا ويتم المروة الخلق ويقتل الناس بعضهم عن بعض ويكون بدو
ذلك من ناحية القبلة وبه تردد ثم يتوجه الى ناحية
المغرب ويفشو في الارضين ويلقى الناس من الغرق بكثرة
المياه شدة ويفسد الاقاليم كلها واول الفساد يكون في الاقليم
الخامس والثالث ويهلك ملوك في السنة الرابعة من هذا
القران ويكون في تلك السنة منازعة ويسفك دما كثيرة
حتى لا يبقى من اصحاب الحروب كبير احد ∴ ويقوى اهل بيت
النبوة والعلم ويتبين الخير من الشر كما يبين الابيض من
الاسود ويخرج في الناس النشو الصالح ويموت ناحيه

223v

وذكر بعض الحساب وهو عبد الله بن محمد بن بشر المنجم انه
امتحن الزهرة فوجدها في الثور في ٥ كه لـ وان ماشا الله
سها في بابها لانه كان يجب ان يريد تعديلها وهو لـ كه لـ على
موضع الشمس بسقفه منه وانه امتحن الطالع بان زاد على طالع
القران الماضي ما كان ماشا الله يزيد من الارج لكل سنه وهو
ثلث وتسعون درجه وخمس عشره دقيقه لتسع عشره سنه
فخرج طالع هذا القران الميزان كه ٥ وهذا على عرضنا نحن

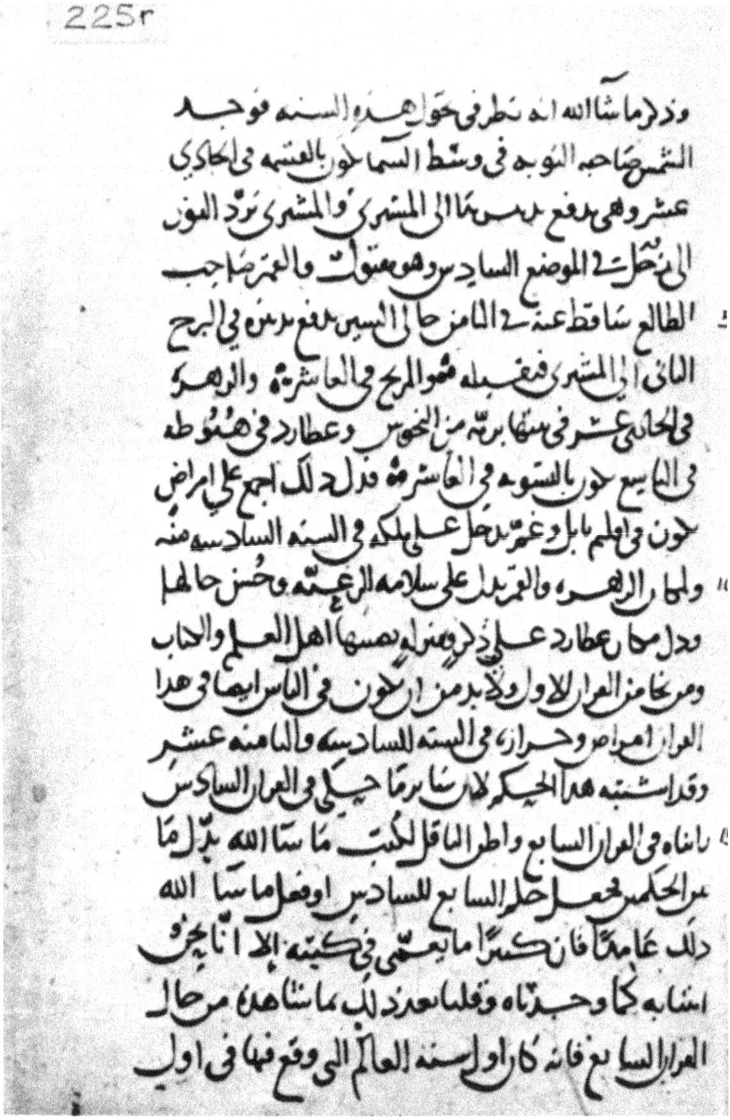

225r

وذكر ماشاء الله انه نظر في تحويل هذه السنة فوجد
الشمس صاحبة النوبة في وسط السماء تكون بالقسمة في الحادي
عشر وهي تدفع تدبيرها الى المشتري والمشتري يرد النور
الى زحل في الموضع السادس وهو منكوس والقمر صاحب
الطالع ساقط عنه في الثامن حال الى الشمس يدفع تدبيره في البرج
الثاني الى المشتري فيقبله فهو المريخ في العاشرة والزهرة
في الحادي عشر في شرفها بريئة من النحوس وعطارد في هبوطه
في التاسع تكون بالسوية في العاشرة فذلك اجمع على امراض
تكون في اقليم بابل وغمّ يحل على ملكه في السنة السادسة منه
ولمكان الزهرة والقمر يدل على سلامة الرعية وحسن حالهم
ويدل مكان عطارد على ذكر منزلة يصيبها اهل العلم والكتاب
ومن كان من القران الاول ولا بد من ان تكون في الناس ايضا في هذا
القران امراض وحرارة في السنة السادسة والثامنة عشر
وقد اشتبه هذا الحكم لانا ربما جعلنا في القران السادس
ما نراه في القران السابع واظن الناقل للكتب ما شاء الله بدّل ما
من الحكمين فجعل حكم السابع للسادس او فعل ما شاء الله
ذلك عامدا فان كثيرا ما يعمي في كتبه الا انا نحن
اشباه كما وجدناه وقلنا بعد ذلك بما شاهدنا من حال
القران السابع فانه كان اول سنة العالم التي وقع فيها في اول

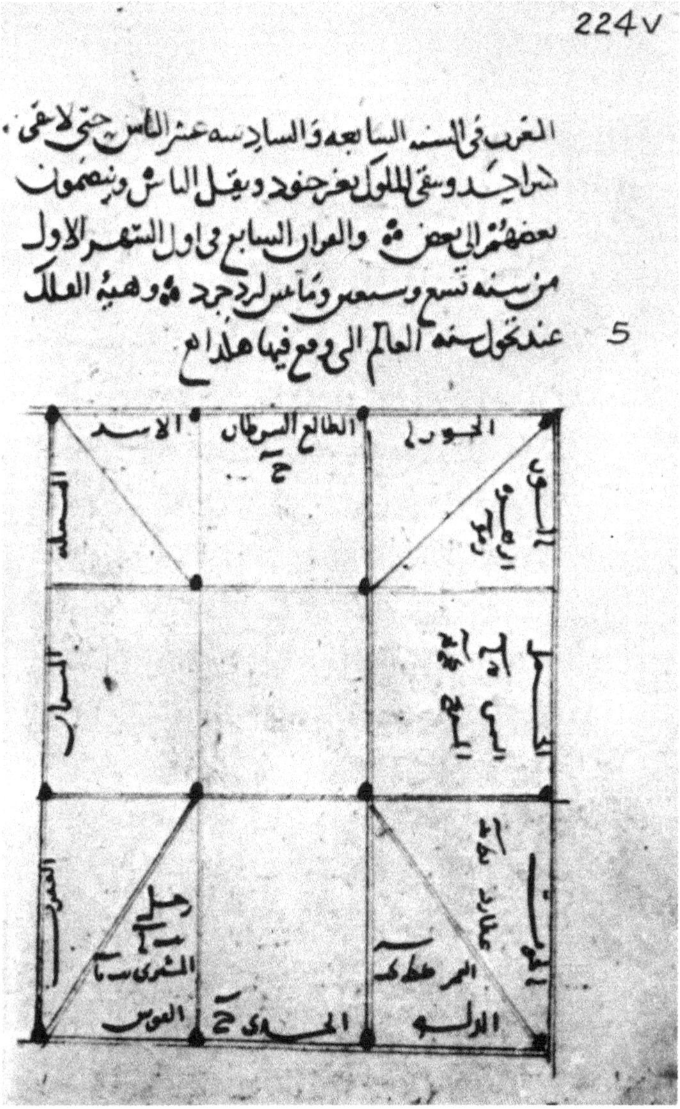

224v

الحرب في السنه السابعه والسادسه عشر الناس حتى لا يخفى
شر احد وسقى الملوك بعز جنود ويقتل الناس وينضمون
بعضهم الى بعض ٭ والقران السابع في اول الشهر الاول
من سنه تسع وسبعين وثمانين ليزدجرد ٭ وهيئة الفلك
عند تحول سنه العالم التي وقع فيها هذا الامر

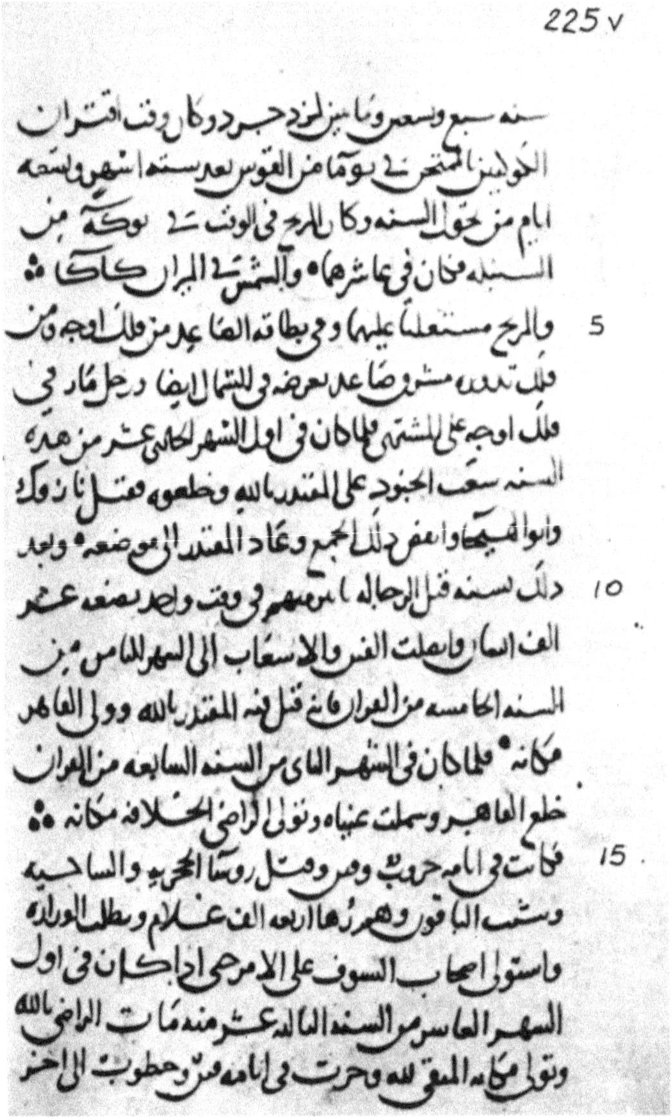

225 v

سنة سبع وتسعين وماتين ليزدجرد وكان وقت اقتران
العلويين المتحيرين في يوما من القوس بعد ستة اشهر وستة
ايام من تحويل السنة وكان المريخ في الوقت في ٢ من
السنبلة فكان في عاشرهما والشمس في الميزان كما كان
والمريخ مستعليا عليهما وفي بطالة الصاعد من ذلك اوجب ومن
ذلك تحدث مشرق صاعد يعرض في الشمال ايضا دخل ما في
ذلك اوجب على المشتري فلما كان في اول الشهر الحادي عشر من هذه
السنة شغب الجنود على المقتدر بالله وخلعوه وقتل نازوك
وابو الهيجاء ثم فرق ذلك الجمع وعاد المقتدر الى موضعه وبعد
ذلك بسنة قتل الرجالة ابن عمهم في وقت واحد نيفا عشر
الف انسان واتصلت الفتن والاشغاب الى الشهر الخامس من
السنة الخامسة من القران فانه قتل فيه المقتدر بالله وولي القاهر
مكانه فلما كان في الشهر الثاني من السنة السابعة من القران
خلع القاهر وسملت عيناه وتولى الراضي الخلافة مكانه
فكانت في ايامه حروب وفتن وقتل رؤسا الحجرية والساجية
وشغب الباقون وهم زهاء اربعة الف غلام وبطلت الوزارة
واستولى اصحاب السيوف على الامر حتى اذا كان في اول
الشهر العاشر من السنة الثالثة عشر منه مات الراضي بالله
وتولى مكانه المتقي لله وجرت في ايامه فتن وحروب الى اخر

2. TRANSLATION

⟦f 214v, 1⟧ In the name of God, the Merciful, the Compassionate; Lord, make smooth the way by Thy mercy. ⟦2⟧ Praise unto God, the Praised, the Glorious, the Creator, and Raiser of the dead, the Doer of whatever He desires, the Maker of ⟦3⟧ night and day, the Knower of what is on land and sea, the Discoverer of the hidden in science ⟦4⟧ and secrets as an indication to His worshipers of His grandeur, and guidance for them unto His wisdom, ⟦5⟧ and an impulsion for them toward the worship of Him alone, there being none associated with Him and nothing to complete Him in any thing. He is the Creator of ⟦6⟧ every thing, and He is the Mighty (one) the Wise (one).

However, after (the foregoing), verily there have preceded (something) of our sayings and what ⟦7⟧ we have collected of science in the art of astrology in the first two parts of this our book as to ⟦8⟧ what has passed. (Now) we mention in this part what Māshā'allāh arranged concerning ⟦9⟧ the times of conjunctions occurring between Saturn and Jupiter, and their transfer (*or* shift) among the triplicities, ⟦10⟧ and what he set forth of the form of heaven at the famous times among them and recorded events ⟦11⟧ which occurred in them involving the pure prophecies and the dazzling miracles ⟦12⟧ and what he judged about it for all of these things and (what) he disclosed as to its evidence, and what ⟦13⟧ Hermes and others mentioned concerning judgments based on conjunctions and world-year transfers (that is, vernal-equinox horoscopes), ⟦14⟧ as to what the ancients had to say concerning judgments on nativity years and so on as to what follows it of the duration (of the reigns) of ⟦15⟧ kings and governors, and the times of wars, and mentioning the condition of prices ⟦16⟧ (*illegible*) with the assistance of God and the goodness of His help. Māshā'allāh said in his book ⟦17⟧ on conjunctions that the first conjunction between Saturn and Jupiter occurred at the beginning of the motion, ⟦18⟧ five hundred and nine years, two months, and twenty-four days having passed of the ⟦19⟧ thousand of Mars. It took place at seven degrees and forty-two minutes ⟦f 215r, 1⟧ of Taurus. He mentions after it the conjunctions that occurred in this earthy triplicity, (one) conjunction after another, ⟦2⟧ adding, for the degrees of the position of each conjunction, to the position of the conjunction preceding it, two degrees ⟦3⟧ and twenty-five minutes, without explaining the defect in (*or* the reason for) that. Then the conjunctions are in one ⟦4⟧ triplicity after another, the duration of each triplicity being nineteen years, ten months, and eleven ⟦5⟧ days until it ended with the transfer of the transit from Libra and its triplicity to Scorpio ⟦6⟧ and its triplicity after two thousand four hundred and twelve years, six months, and twenty-six (*marginal note:* 2412 years, 206 days) ⟦7⟧ days, this being the one in which the Deluge occurred. The first of the conjunctions of ⟦8⟧ this transit was in (one) degree and twenty-four minutes of Scorpio, and its duration (was) ⟦9⟧ nineteen years, ten months, and ten days, and the form of the heavens at ⟦10⟧ the entry of the sun into Aries for the year in which this first conjunction of ⟦11⟧ the transfer of the transit took place, according to what Māshā'allāh says is this picture: ⟦f 215v⟧ [*Horoscope 1*]

⟦1⟧ Māshā'allāh said: No planet except Jupiter can be found for the guardianship of this year because it is ⟦2⟧ the lord of the ascendant, and the lord of its triplicity is the sun alone and it confers its power upon it, ⟦3⟧ and (it receives power) from Venus because she (Venus) is the lord of the domicile of the moon, and she is in her exaltation, and it (the moon) is in the ⟦4⟧ sixth place displaced from its position, and Saturn is in the twelfth, retrograde, and Jupiter and Mars ⟦5⟧ in the eleventh, they will be by division (*qisma*) in the tenth, both retrograde. The retrogression of ⟦6⟧ the three planets indicates, at the transfer of this transit in upper midheaven and a human (*lit.* spiritual) sign, the spiritual ⟦7⟧ death of the earth in the second conjunction of this transit because the horoscope (*here some folios of the MS have been bound out of order*) ⟦f 226r, 1⟧ (is) a sign having two bodies, so look at the shift of the transit from one triplicity to another. ⟦2⟧ And if you see the transit

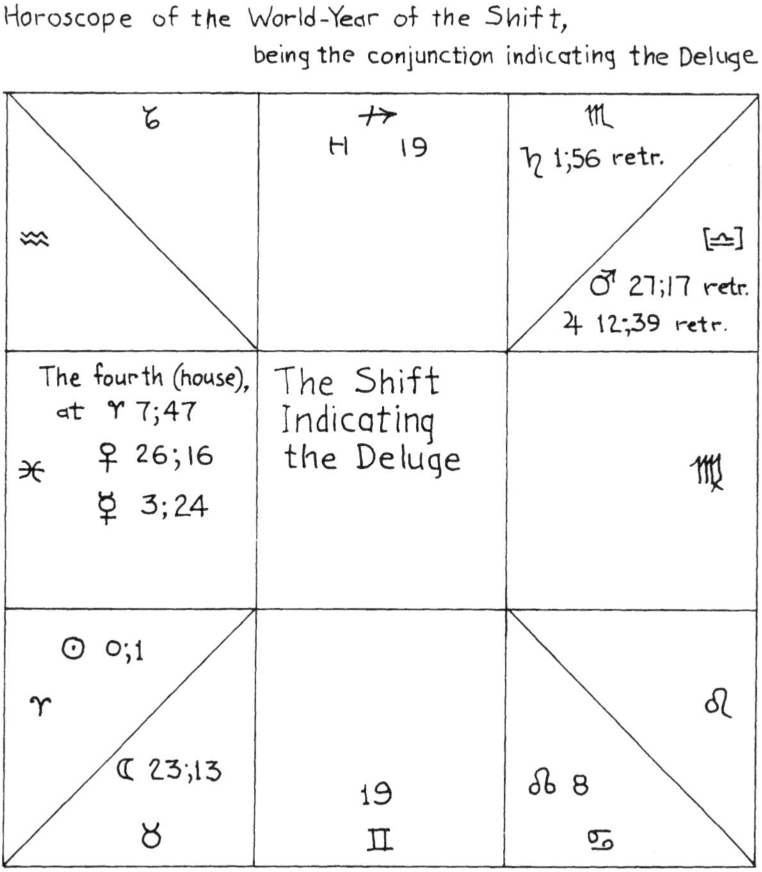

Horoscope 1.

informing ahead of miracles, or death, or nativities of prophets, then consider the ascendant of ⟦3⟧ the year of the shift. If it has two bodies, especially Sagittarius, there will not be ⟦4⟧ anything at the first conjunction, and from the second conjunction the news of the event is made known. If ⟦5⟧ the sign is (either) fixed or moving, then the time of the event will be at the first conjunction from ⟦6⟧ the shift of the transit. If the moon, it being the lord of the light of the night in this ⟦7⟧ year, had been spoiled, nothing would have remained of the world, but its (the moon's) being safe indicates growth and increase (in numbers) ⟦8⟧ after death. The occurrence of the second conjunction of this transit in which was the ⟦9⟧ Deluge was in

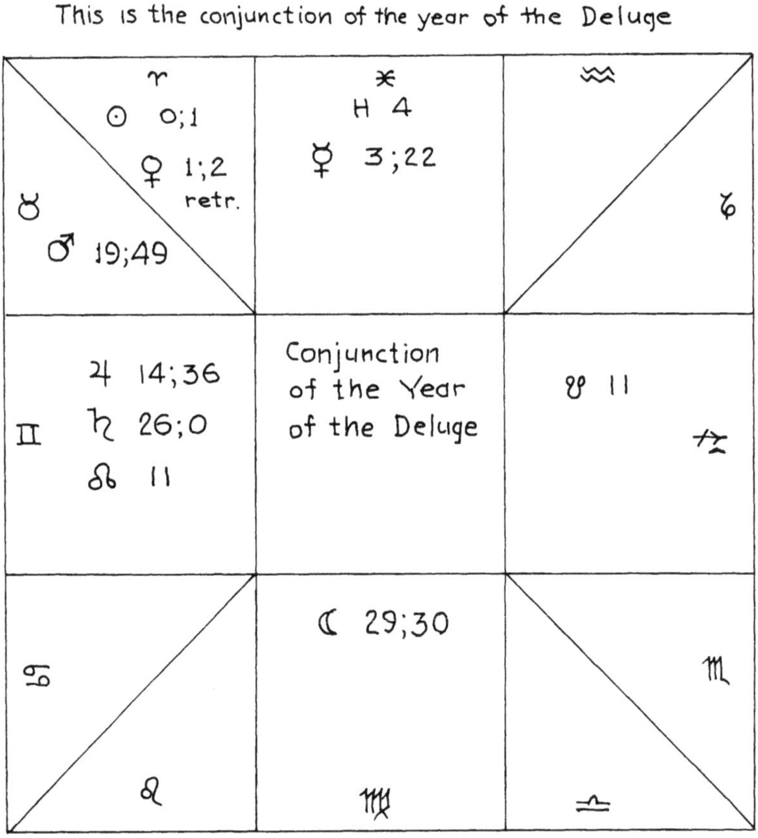

Horoscope 2.

Cancer in three degrees and forty minutes, and the form ⟦10⟧ of the heavens at the time of entry of the sun into Aries for the year in which it occurred is according to what ⟦11⟧ is established in this picture according to the way Māshā'allāh described of that, and the ⟦12⟧ Deluge occurred in its fifth month.
⟦f 226v⟧ [*Horoscope 2*]

⟦1⟧ Māshā'allāh said: I looked at the time of the death of the earth which is indicated by the retrogression ⟦2⟧ of the superior planets at the time of the shift of the transit to the watery triplicity (and realized) that it would be in ⟦3⟧ its second conjunction. So I found Jupiter to be the best planet for the guardianship of this year, because it is the ⟦4⟧ lord of the ascendant, and the lord of its term and the triplicity of the sun, and it (the sun) and Venus, ⟦5⟧ the lord of the exaltation of the ascendant, conjoin, ⟨and aspect Jupiter⟩ and it (Jupiter) is in the cardine of the earth in a sign which does not ⟦f 227r, 1⟧ fit it, together with Saturn. If it were not together with it (Saturn) in it (the cardine of the earth) and the lord of its house in the ascendant, ⟦2⟧ that would indicate that what was previously described concerning death will be in this year, ⟦3⟧ in its fifth month, at the eastern apparition of Saturn. Now Jupiter is in Cancer, ⟦4⟧ and if it were not for the place of Jupiter, and the lord of its house, which is the moon, whose turn (?) is the night, is in the cardines, ⟦5⟧ the event would have occurred in the fifth year of the conjunction. But their place being in the cardines ⟦6⟧ it (the event) was hastened, which put it in the fifth month of its first year. So let this be understood, ⟦7⟧ and know that if the superior planets, which are above the sun, are retrograde at (the time of) ⟦8⟧ the transfer of the shift of the transit from one triplicity to another, or at the year-transfer of one of the conjunctions, ⟦9⟧ (this) indicates death by heavenly misfortunes. But if the two planets ⟦10⟧ inferior to the sun are retrograde in it, those misfortunes will be terrestrial. But if all of them ⟦11⟧ are retrograde, the misfortunes will be celestial and terrestrial also, so watch out. But ⟦12⟧ if they are in spiritual, human signs, they being Gemini and Virgo and Libra and the first half ⟦13⟧ of Sagittarius and all of Aquarius, the misfortunes will involve people. But if they are in ⟦14⟧ other signs, animal signs, the misfortunes will involve the kinds (of animals) which are associated with ⟦15⟧ these signs. So if you want to know the kind of misfortune and torment, ⟦16⟧ look at the shift of the conjunctions from one triplicity to another unto the lot of the transit, in whatever ⟦17⟧ sign it takes place, and if it takes place in a fiery sign, the misfortunes would be from fire. But (if) in a watery sign, ⟦18⟧ the misfortune will be from water, and if (in) an airy sign the torment will be from wind, and if in an ⟦19⟧ earthy sign, it will be from a landslide or earthquake or falling rocks. And what will be from ⟦f 227v, 1⟧ these indications in the transit indicates the mighty things which occur in the conjunctions, ⟦2⟧ and what will be of them at the time of the conjunction indicates what will happen in its year. ⟦3⟧ What is of them at the time of the year-transfer of the world-years (that is, vernal equinoxes) indicates what will happen ⟦4⟧ in its months. Each of these stages is easier than what preceded it. ⟦5⟧ So between the

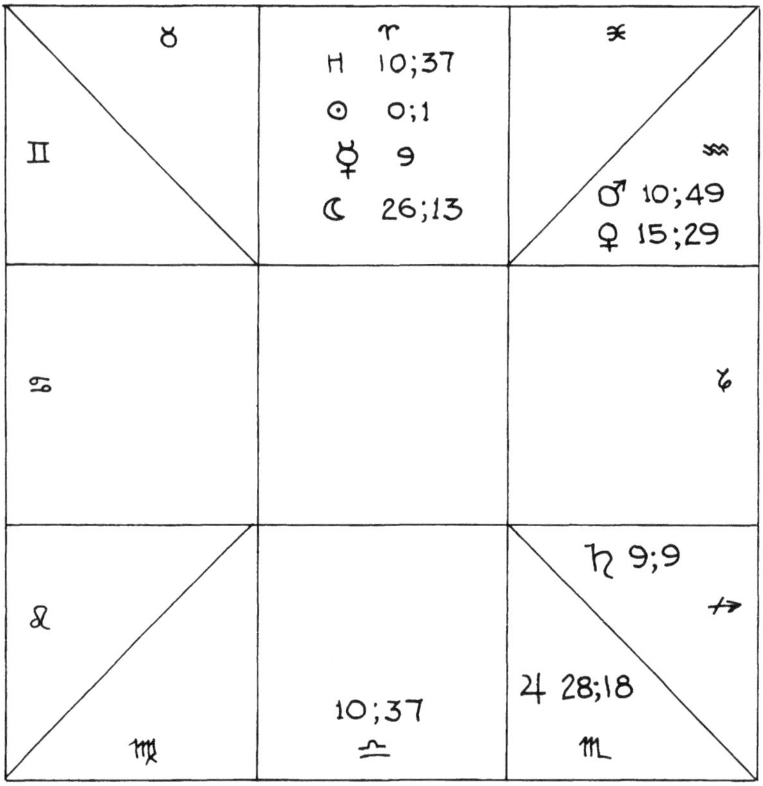

Horoscope 3.

time of the first conjunction of Saturn and Jupiter at the beginning of the motion and ⟦6⟧ the month of the Deluge, after the placing (aside) of the nine years, two months, and twenty-four days ⟦7⟧ increased over the five hundred years which Māshā'allāh mentioned as having passed of the thousand of Mars, ⟦8⟧ if Māshā'allāh had performed it (correctly) by computation, the method being to put ⟦9⟧ two thousand, four hundred, and twenty-three years, six months, and twelve days, ⟦10⟧ and Māshā'allāh arranged the conjunctions after that, triplicity by triplicity, until he arrived ⟦11⟧ at the fiery triplicity in which was the birth of the Anointed (Christ), upon Him peace, in the ninth conjunction of them, ⟦12⟧ it being the triplicity of Leo, so the eighth conjunction of this transit came to be in Sagittarius, ⟦13⟧ eighteen degrees and fifty-five minutes, and its time nineteen years, ten ⟦14⟧ months, and ten days, and the first of it was at the end of five thousand, seven hundred, and forty-nine years, ⟦15⟧ five months, and eight days from the beginning of the conjunction. The form of ⟦16⟧ the heaven at the entry of the sun into Aries at the time of the year-transfer of the world-year in which it took place ⟦17⟧ is according to what is put down in this picture. ⟦f 228r⟧, [*Horoscope 3*]

⟦1⟧ And Māshā'allāh explained that when he looked at it he could find no planet stronger than the sun, because it is in ⟦2⟧ the ascendant and it is the lord of its exaltation. It confers its counsel upon Saturn, and it (Saturn) is in the ninth, the position of ⟦3⟧ prophecy, and so (also) are the moon and Mercury, so that a prophet will be born; God will illuminate darkness with him, ⟦4⟧ and give sight to the blind. And because Saturn, which receives the strength of the two luminaries from the ascendant, is ⟦5⟧ in Sagittarius, it indicates that his birth will be in the second conjunction from this (first) conjunction, ⟦6⟧ and because of the aspect in trine, it indicates compassion and gentleness, and because of the locus of Saturn with respect to the lord ⟦f 228v, 1⟧ of its house [*read* sign], it indicates what violence befalls him from his people. The ninth conjunction, ⟦2⟧ in the thirteenth year of which the Christ was born, was in eleven degrees and twenty ⟦3⟧ minutes of Leo, and the form of the heavens at the entry of the sun into Aries for the year in which it ⟦4⟧ occurred, is according to this picture [*Horoscope 4*]

⟦5⟧ And Māshā'allāh explained that when he looked at this picture he found the sun, the luminary of the day, ⟦f229r, 1⟧ and Mercury, the lord of the house of the moon, the two of them sending their counsel unto Saturn, the lord ⟦2⟧ of the seventh and the eighth from the ascendant, it being accepted in its position, increasing the forward motion, ⟦3⟧ so that indicates that the prophet whose birth was indicated by the first conjunction (will be) ⟦4⟧ born in this conjunction at the end of the thirteenth year. And verily he found the form of the heaven ⟦5⟧ at the time of the year-transfer of the world-year in which the Christ, upon Him peace, was born, it being ⟦6⟧ the thirteenth, according to this picture [*Horoscope 5*].

⟦f 229v, 1⟧ When he looked at the places of its planets, he found the lord of the year to be Saturn, because of its strength and its numerous ⟦2⟧ aspects, and because the planets, all of them except Mars and the moon, were conferring their counsel upon it, ⟦3⟧ it being in its house, and the ascendant its exalta-

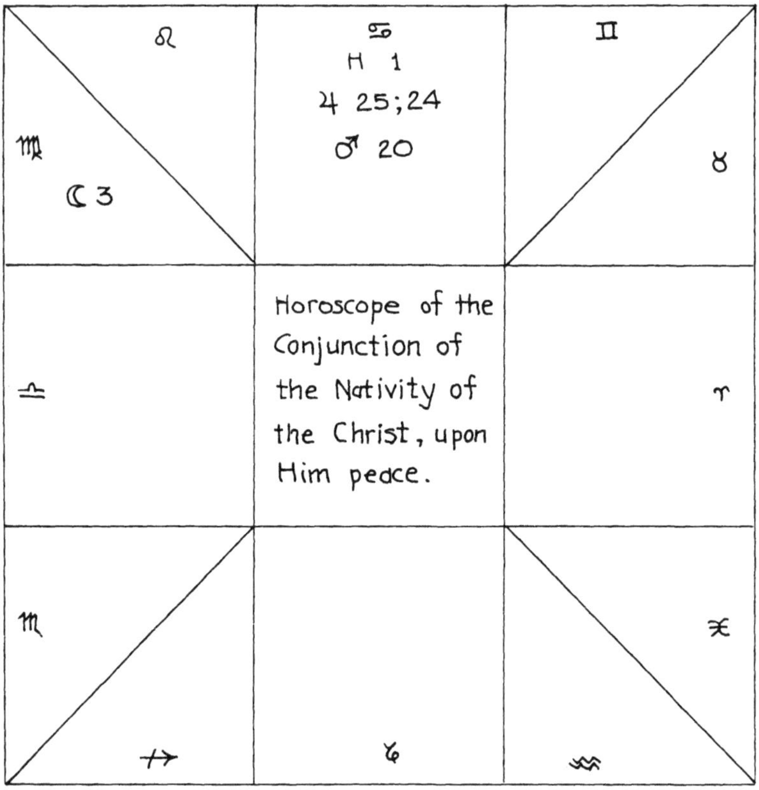

Horoscope 4.

tion and triplicity, and the lord of the ascendant confers its counsel ⟦4⟧ upon it from its second house. That indicates that he will be born in ten months of this year, ⟦5⟧ and because the transfer is nocturnal and the moon is with Mars in the ninth, it indicates violence he will meet, and fear ⟦6⟧ about him of being killed in a high place, because the two are conjunct in the highest of the signs. So there will be ⟦7⟧ between the time of the first conjunction, at the beginning of the motion and the year in which Christ, upon Him peace, was born, ⟦8⟧ it being the thirteenth (year) of this conjunction, after the dropping of the nine years and months by which it was in excess of ⟦9⟧ the five hundred years passed of the thousand of Mars – for Māshā'allāh entered them ⟦10⟧ in the computation (the way to do it is to not enter) – five thousand, seven hundred, and seventy-three years, and twenty-four days (*gloss in a different*

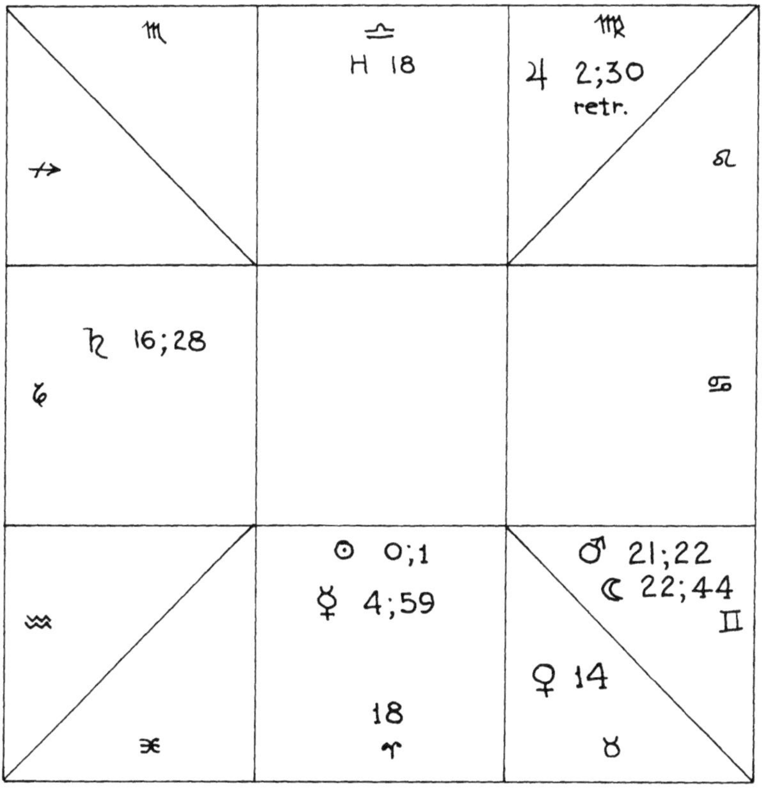

Horoscope 5.

hand: Between the Deluge and the birth of Christ, upon Him and upon *our* Prophet peace, (was) three thousand, three hundred, and forty-nine solar years), ⟦11⟧ and between the time of the Deluge and the end of ⟦12⟧ the year in which the Christ was born (was) three thousand, three hundred, and forty-nine solar ⟦13⟧ years, six months and twelve days. And the conjunctions continued after that ⟦14⟧ in their succession until the transit was carried from the airy triplicity, which is Gemini and what belongs to it, ⟦15⟧ unto Cancer and its triplicity in which was the state of the Arabs at the end of ⟦16⟧ six thousand, three hundred, and forty-five years, three months, and twenty days, in which we put ⟦17⟧ the nine years and months which are in excess of the five hundred years passed of the thousand of Mars, ⟦18⟧ so there remain six thousand, three hundred, and thirty-six years, and twenty-six days from ⟦19⟧ the time of the conjunction at the beginning of the motion. The first of the conjunctions of this ⟦f 230r, 1⟧ triplicity took place at Scorpio four degrees and two minutes, it being the conjuction of ⟦2⟧ the religion. In its second year was born the Prophet Muḥammad, the prayers of God upon him and peace. (*gloss:* The conjunction indicating the religion of Islam.) ⟦3⟧ Māshā'allāh said, and others than he of the learned, that the transfer of the world-year in which ⟦4⟧ this conjunction occurred was at the second hour of the night the morning of which was ⟦5⟧ twenty-three Bahman, but there is no date except that previously mentioned from the beginning of ⟦6⟧ the motion, because there there were no years (given) to use (as a base). And the form of the heaven was according ⟦7⟧ to this picture [*Horoscope 6*].

⟦f 230v, 1⟧ Verily it (the divisor) reached Libra from the conjunction of the Deluge, and from the sign of the *intihā'* of the cycle to ⟦2⟧ Gemini. So from here began the cycle of the Arabs, for Venus and Mercury, of which this ⟦3⟧ year is its first, and its duration is three hundred and sixty years. And from the position of the Saturn-Jupiter conjunction ⟦4⟧ at the time of the Deluge (the divisor reached) unto Sagittarius, and from the division (*qisma*), which allocates one degree ⟦5⟧ per year, (came) up to the end of nineteen degrees of Pisces and the beginning of the twentieth degree ⟦6⟧ of Pisces, the beginning of each cycle. So there is between the entry of the sun into Aries in ⟦7⟧ this year and the first day of the year of the Hijra, in Persian years ⟦8⟧ fifty-one years, three months, eight days, and sixteen hours. And from the ⟦9⟧ first day of the Hijra until the day of the kingship of Yazdijird the king is nine years, eleven ⟦10⟧ months, and nine days Persian. The total of that will be sixty-one years, ⟦11⟧ two months, seventeen days, and sixteen hours. So there was between ⟦12⟧ the time of the Deluge until the first conjunction of the religion three thousand, nine hundred, and twelve ⟦13⟧ years, six months, and fourteen days. And from the time of ⟦14⟧ the Deluge until the first of the rule of Yazdijird was three thousand, nine hundred, and seventy-four ⟦15⟧ years, nine months, and one day. But Abū Ma'shar mentioned that it was ⟦16⟧ three thousand, seven hundred, and thirty-three years, two months, twenty-

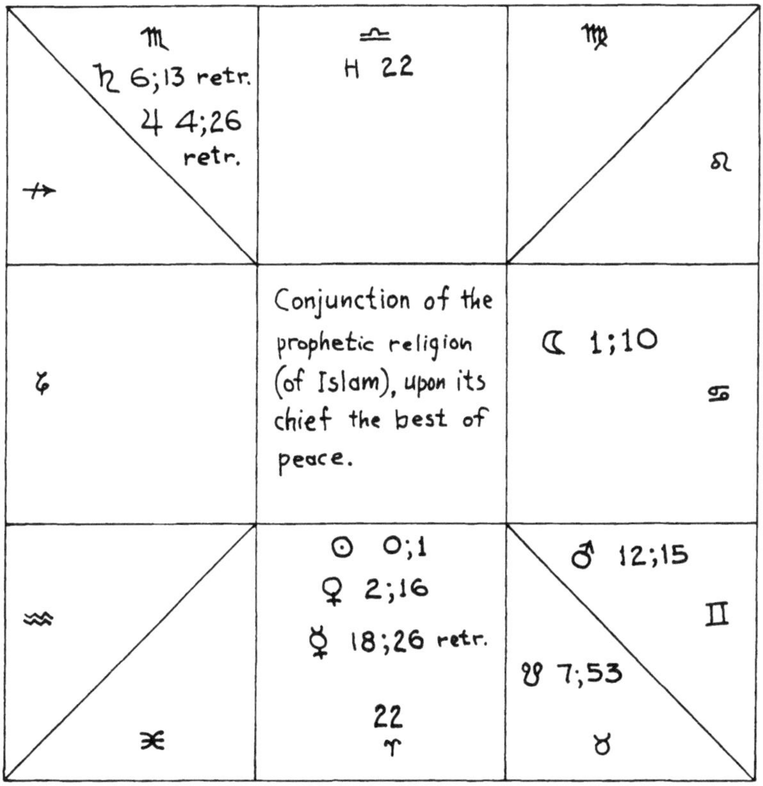

Horoscope 6.

six ⟦17⟧ days, nine minutes, and ten seconds. Māshā'allāh said that when he looked at ⟦18⟧ this picture and the position of its planets, he found the strongest of them and the strongest in witnesses to be ⟦19⟧ Saturn, because it entered the ascendant and it is the lord of its exaltation; and the moon is lord of the light of ⟦f 231r, 1⟧ the night, conferring its counsel upon it (Saturn) from the tenth by counting (*'adad*) and (this is) the ninth by division (*qisma*), the place of prophecy, ⟦2⟧ and Jupiter confers counsel upon it (the moon). Mars is found in Gemini, it being the ninth by counting ⟦3⟧ in the highest (part) of the orb (*minṭaqa*). ascending to its extreme. It accepts the counsel of the sun, Venus, and Mercury, ⟦4⟧ (which is) retrograding. All that, plus the place of the first lot, it being Saturn in Scorpio, and its lord ⟦5⟧ in the place of religion, indicate that there would be born in the second year of this conjunction a prophet in Tihāma, ⟦6⟧ the land of the Arabs, and that is because the place of the first lot is in Scorpio, and (because of) the connection of the moon with Venus. ⟦7⟧ If the ascendant were a sign having two bodies and Saturn were in a sign having two bodies, I would have said that his birth ⟦8⟧ would be in the second conjunction because of the reception of Venus and Mars in their places. The place of the moon in ⟦9⟧ upper midheaven indicates that he will be secure from being killed, and because Venus escaped combustion ⟦10⟧ (he means [its] emergence from the ray), this indicates that he will meet hardships and go into hiding for a while; then he will establish connections and ⟦11⟧ be strengthened, and become well known, and the people of his doctrine will rule. Because the place of the moon is in upper midheaven and Mars is in the house of ⟦12⟧ religion (these) indicate that he will seek religion and its being established by struggle. And because of the connection of the moon with Venus ⟦13⟧ the people of his doctrine will have the nature and manners of Venus. The people of Tihāma, in the second year ⟦14⟧ of this conjunction, will meet hardships from enemies until they flee and are dispersed in the cities. ⟦15⟧ It will be thus because this year is the Year of the Elephant, and it is it in which is born the Prophet, the prayers of God upon him and peace. ⟦16⟧ Then there pass the years of the first conjunction, they being nineteen years, ten months, and ten days, ⟦17⟧ and the years of the second and the third conjunctions, and all that amounts to fifty-nine ⟦18⟧ years and seven months, the end of which was before the time of Yazdijird's rule by one year, ⟦19⟧ nine months, seventeen days, and sixteen hours. And there befell the ⟦f 231v, 1⟧ fourth conjunction in Scorpio eleven degrees and nineteen minutes. ⟦2⟧ Māshā'allāh said that the form of the heaven at the entry of the sun into Aries the year in which ⟦3⟧ this conjunction befell (was) according to this picture [*Horoscope* 7]. ⟦4⟧ So he looked at its ascendant and the positions of its planets when the conjunction returned to the base sign ⟦5⟧ which indicated the birth of the Prophet, the prayers of God upon him and upon his family and peace. So the positions of the two maleficent (planets) ⟦6⟧ in the two cardines, and the sun and moon and Jupiter being cadent

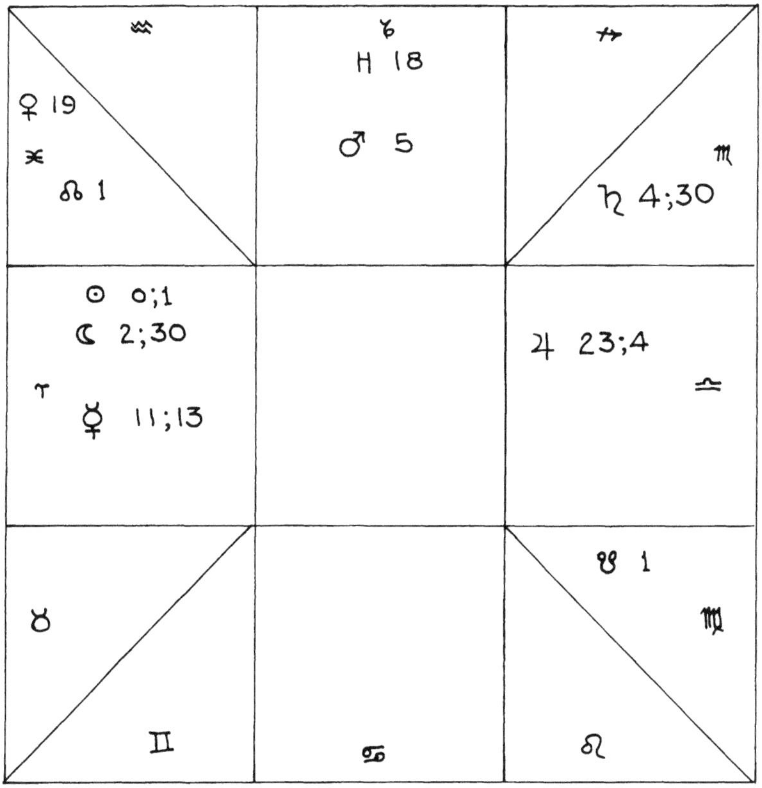

Horoscope 7.

from the cardines, and the strength of Saturn ⟦f 232r 1⟧ in this conjunction and numerous aspects and its being in the sign of Scorpio (*from margin:* indicate the strength of the Arabs) in it and their victory ⟦2⟧ over those who oppose them. And the conferring of its (Mars') counsel upon it by Mars, it being the lord of the exaltation of the ascendant (*margin:* that is, in this conjunction), ⟦3⟧ and the reception of one of the two by the other, indicate disputes and the shedding of much blood and the death of the king of the ⟦4⟧ fourth climate and the victory of evil people over the king. And because Mars is in the ascendant it indicates the going of ⟦5⟧ its king to the East and his death there. And because Saturn is in upper midheaven it indicates general ⟦6⟧ misery and violence, and because it is in the sign of the Arabs, and Venus is with the ascending node in its exaltation, it indicates ⟦7⟧ the strength of the Arabs and their seeking the kingdom. And because of the position of Saturn with respect to the horoscope, ⟦8⟧ which is its house, (it) indicates fear of the people of the house of the Prophet, the prayers of God upon him and peace, in the three ⟦9⟧ years passing of this conjunction. And because the equinox is nocturnal and the moon is under ⟦10⟧ the ray and the sun unfortunate and both of them cadent from the fourth entering ⟦11⟧ the third, it indicates what will assail the people of Babylon of violence, and will disperse them in the cities, and the victory of ⟦12⟧ their enemies over them and their king. This conjunction is the one in which the Arabs were victorious over ⟦13⟧ the king of the Persians and his country and their making him depart from it, and their ruling it without him. The expulsion of Yazdijird ⟦14⟧ from al-Madā'in was after there had passed three years of it, and at the end of four years and a fraction ⟦15⟧ of his rule. Then after it was the fifth conjunction, and its duration was nineteen years, ⟦16⟧ ten months, and eleven days, in Cancer thirteen degrees and forty-two ⟦17⟧ minutes, and the form of the heaven at the year-transfer of the world-year which occurred ⟦18⟧ at it is according to this picture ⟦f 232v⟧ [*Horoscope 8*]. ⟦f 233r⟧ (This folio is blank in the MS.)

⟦f 233v, 1⟧ Māshā'allāh said that when he looked at this transfer, at the ascendant and the positions of the planets, ⟦2⟧ the sun was the lord of this year and it confers its counsel upon Saturn from quartile (aspect). ⟦3⟧ So that indicates the death of the king of Babylon and his elimination in the fourth year of it, and ⟦4⟧ the occurrence of disputes at the end of the ninth year, and corruption and much shedding of blood ⟦5⟧ because Saturn, which is the counselor of the sun in quartile, comes to be in opposition to it in its (Saturn's) exaltation, and disputes the kingship with it (the sun). ⟦6⟧ If the sun did not confer its counsel upon Saturn, it (the sun) being the lord of the year, it would not have indicated, ⟦7⟧ after the elimination of the king of Persia, fighting and wars; nevertheless, it conferred ⟦8⟧ counsel upon it, so when it (Saturn) came to be in its exaltation in opposition to it (the sun), it disputed the kingship with it (the sun), and so it indicates what I mentioned, ⟦9⟧ and it does not cease to be thus until it (Saturn) goes out of its exaltation and comes to be in Scorpio, and he (Yazdijird) will die in a region ⟦10⟧ of the East, and the crops will be spoiled in the first and second years. Since the grand con-

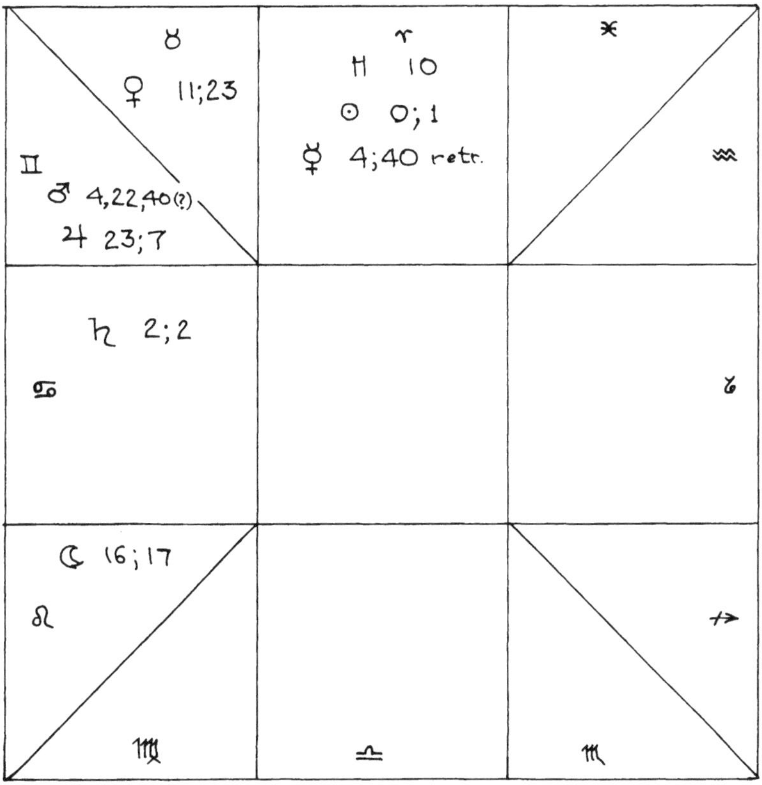

Horoscope 8.

junction was ⟦11⟧ changing from Gemini to Cancer, and from Libra to Scorpio, ⟦12⟧ and from Aquarius to Pisces, it indicates that there will remain of the seed of this dead king ⟦13⟧ and of his family he who was in the land guarded by Pisces, in Ṭabaristān and what adjoins it. ⟦14⟧ And because Jupiter is cadent from the cardine and its position (is) in Gemini with Mars, and because of the corruption of the sun by ⟦15⟧ Saturn and of the moon by the descending node, (these) indicate the death of kings of the people and their notables, and the shedding of ⟦16⟧ much blood on the land of Cancer, which is the fourth climate. ⟦17⟧ After this (is) the sixth conjunction, and its time is nineteen years, ten ⟦18⟧ months, and eleven days. The time of the seventh conjunction is the same, (*end of the inserted folios*) ⟦f 216r, 1⟧ and the eighth conjunction likewise. The ninth conjunction is like it as to time, and ⟦2⟧ all the years of these conjunctions (taken together) are one hundred and nineteen years, two months, ⟦3⟧ and six days. Of these, before there followed Yazdijird the king (there was) one year, nine months, and seventeen days, and the rest are years of Yazdijird, ⟦5⟧ one hundred and seventeen years, six months, and nineteen days, and then (*a gloss has the computation:*

years	months	days
119	2	6
1	9	17
117	4	19

and that was in Rabī' I, A.H. 132) ⟦6⟧ befell the tenth conjunction indicating the Abbasids, and the end ⟦7⟧ of the Ummayads in Scorpio, the sign of the religion. The year-transfer occurred ⟦8⟧ for this conjunction at the end of twelve hours on ⟦9⟧ seventeen Bahman one hundred and nineteen [*read* seventeen] ⟦10⟧ Yazdijird. (*Gloss:* That is six Sha'bān, A.H. 131). The form of the heaven as mentioned by Māshā'allāh ⟦11⟧ is according to this picture ⟦f 216v⟧ [*Horoscope 9*]

⟦1⟧ The [*read* lord of the] fourth year reached, from the conjunction of the religion, Cancer, and from the place ⟦2⟧ of the conjunction, Leo, and from the degree of the division (*qisma*) put at the beginning of each cycle ⟦3⟧ at twenty degrees of Pisces (it reached) nineteen degrees of Virgo. ⟦4⟧ In it is the ray of Mars, and of Mercury, which is retrograde. Verily Māshā'allāh explained ⟦5⟧ that he looked at the transfer of this year in which the conjunction returned ⟦f 217r, 1⟧ to Scorpio, it being the sign which indicated at the beginning the birth ⟦2⟧ of the Prophet, upon him the prayers of God and peace. Its lord was in the fifth with ⟦3⟧ Mercury, which is retrograde, the two not being in aspect in trine with Saturn, and that indicates ⟦4⟧ pestilence among the people from plagues at the beginning of this year. ⟦5⟧ Because of the conferring by Venus of its counsel upon Saturn, it being with Jupiter in Scorpio ⟦6⟧ retrograde, this indicates mighty wars in which there will be much slaughter in the East and ⟦7⟧ degeneracy and spoiling and the disappearance of their piety. And because of the conferring by the lord of the ⟦8⟧ ascendant and by the lord of the ninth, and the lord of the fourth of their counsel upon Saturn, it ⟦9⟧ being in the ascendant, this

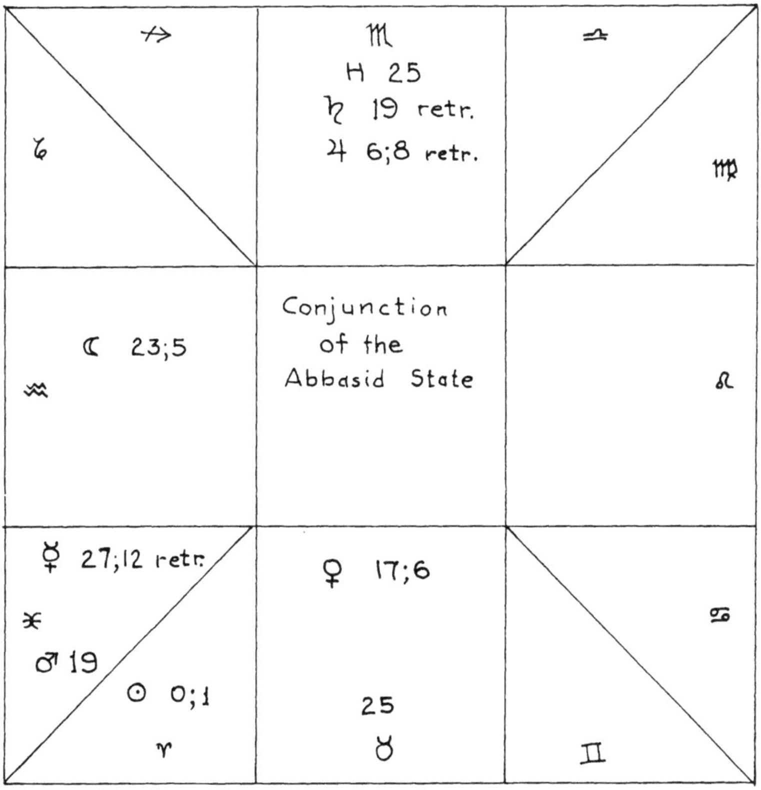

Horoscope 9.

indicates that the people will become subject to the people of the East ⟦10⟧ and the government will be in their hands, and because the place of the conjunction is in Scorpio, which is the sign of ⟦11⟧ the Arabs, and the corruption of Venus, it being their planet, this indicates their death by plagues and ⟦12⟧ wars. I judged calamity and violence for two reasons. Because Mars, ⟦13⟧ being the lord of the house, is in a sign having two bodies, and there will be also in the fifth year ⟦14⟧ a war and shedding of blood in the western region, and during it people of the East will be killed, ⟦15⟧ many people, because of Mars' place, which is the lord of the ascendant in (a sign indicating) ⟦16⟧ the western region. And because the indication of plagues is Mercury and its place is with Mars, it being spoiled ⟦17⟧ by the retrogression, this indicates plagues, the destruction of cities and castles and the prevalence of this ⟦18⟧ among all the Arabs. All the years of this conjunction will be hard for them ⟦19⟧ because it is in their sign, and is a cardine and Venus, which belongs to them, is unfortunate in another cardine, ⟦f 217v, 1⟧ in opposition to Saturn, which is retrograde. The lord of their upper midheaven ⟦2⟧ is Mercury, it being retrograde and unfortunate, so it indicates violence of ⟦3⟧ many sorts, death and killing and brutality and oppression on the part of the sultan. ⟦4⟧ Then after this tenth conjunction, in which the state and kingdom was transferred from the ⟦5⟧ Ummayads to the Abbasids and there occurred in it wars, and in it ⟦6⟧ blood was shed, and cities were destroyed, was the eleventh conjunction. ⟦7⟧ The twelfth conjunction being the last conjunction of the Scorpio triplicity, all the years of these ⟦8⟧ three conjunctions will be fifty-nine years, seven months, and three days. ⟦9⟧ The last of them will be at the end of one hundred and seventy-seven years, one month, and twenty⟨-two⟩ days ⟦10⟧ of Yazdijird. (*Gloss:* That is in Sha'bān, A.H. 193). The conjunctions were carried to the fiery triplicity of Sagittarius ⟦11⟧ at this time, and the year-transfer in which this first conjunction took place was ⟦12⟧ at seven and a half hours on the twenty-second day of ⟦13⟧ the month of Bahman, one hundred and seventy-seven Yazdijird, and the form of the heaven is ⟦14⟧ according to what I showed in this picture. This year reached Cancer from the conjunction of the ascendant of ⟦15⟧ the religion, and from the place of the conjunction in it (it reached) Leo. ⟦16⟧ And from the division (it reached) nineteen degrees of Scorpio.

⟦f 218r⟧ [*Horoscope 10*] ⟦1⟧ Māshā'allāh said that when he looked at this picture and contemplated its ascendant and the positions of ⟦2⟧ its planets, he found no planet having a better place than Venus because it is ⟦3⟧ the lord of the triplicity of the ascendant in its exaltation in the degrees of upper midheaven ⟦4⟧ and received by the lord of the exaltation of the ascendant and ⟨aspected by⟩ the lord of its house, which is the house ⟦5⟧ it is in. Since the ascendant is of its (Venus') nature, it (Venus) indicates the well-being of the Arabs until the ⟦6⟧ sixth year of the conjunction, then they will be overpowered by the people of their kingdom, who ⟦7⟧ will conspire against them and fight with them, and they (the people) will be helped by the people of the East ⟦f 218v, 1⟧ and Isfahān because of the place of the conjunction in Sagittarius, and the rulership

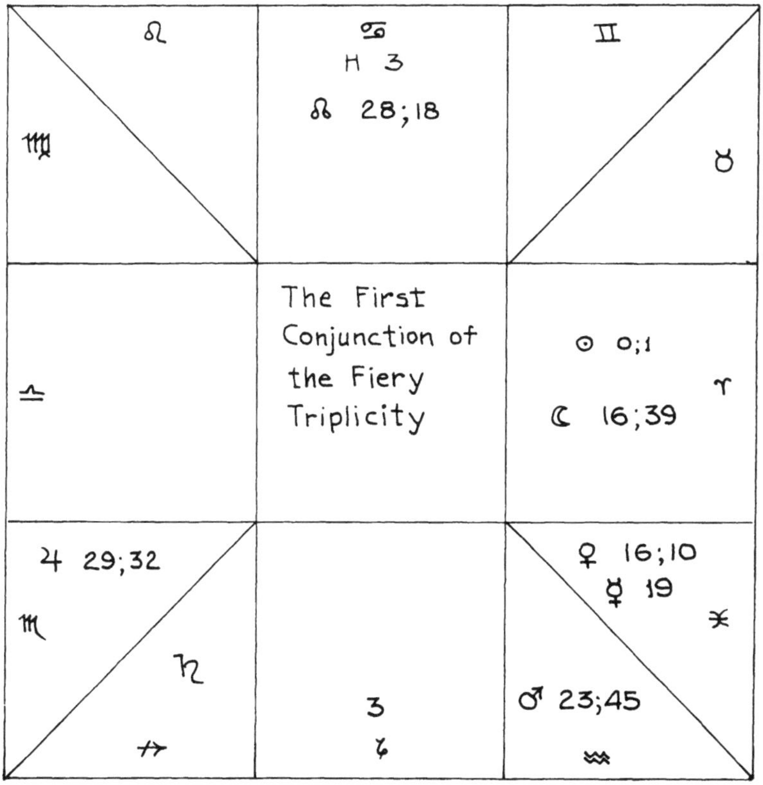

Horoscope 10.

will be transferred from one house to ⟦2⟧ another in the fourth year of the conjunction. Death will increase in the ⟦3⟧ tenth year of it; the government will be transferred to the East because of the transfer of the conjunction ⟦4⟧ to Sagittarius. The years of the kings in this triplicity will be lengthened, and the sword will be lifted up against ⟦5⟧ the people because Mars, it being the lord of the eleventh, was in the eighth ⟦6⟧ in the sign of evil, and the planets were made maleficent by it. If it were not for the reception the matter would have been ⟦7⟧ much worse and more violent. But the reception lightens the evil, and because the ascendant of this ⟦8⟧ conjunction, which is the first of the transfer, fell in Cancer, which is one of the cardines of ⟦9⟧ the ascendant of the first transit, which was in Scorpio and its triplicity, and which indicated the birth of ⟦10⟧ the Prophet, the prayers of

God upon him and peace, and the dominion of the Arabs and the carrying over of the rule ⟦11⟧ unto them, and the house of the sultan is from him, and the house of the moon, and in this conjunction the moon was in ⟦12⟧ upper midheaven, fixed in the cardine, and received in its place by the lord of ⟦13⟧ its domicile, increasing in light, so it indicates the permanence of this state in the hands of its people and ⟦14⟧ their sultan's continuing to be from among them, because Cancer is the ascendant of this shift, it being the house of ⟦15⟧ the sultan, from the past conjunction and the establishment of the state in the hands of the Arabs. ⟦16⟧ The shift from one triplicity to another does not shift from one tribe [? *buṭn*] to another, ⟦17⟧ nor from one people to another, and because it is in an eastern sign the dominion was for ⟦18⟧ the king of the East and victory was for him and the shift unto him. And because the conjunctions rotate in ⟦19⟧ Sagittarius and Leo and Aries, victory was for the Easterners of the people of ⟦f 219r, 1⟧ Khurāsān and the Turks, and those allied to these countries, over the Arabs, ⟦2⟧ without the dominion departing from them, because of the connection we mentioned previously between the conjunctions, ⟦3⟧ which we explained, and it remains thus until the conjunction shifts from ⟦4⟧ the Sagittarius triplicity to another triplicity where there is no mixing or no connection between its horoscope and the cardines of ⟦5⟧ this triplicity. The conjunction of Saturn and Jupiter continues to be ⟦6⟧ in it according to what happened in the days of the Persians at the shift of the conjunctions ⟦7⟧ from the triplicity of Gemini to the triplicity of Scorpio, and there was no relation or mixing between the two transits. ⟦8⟧ So this indicates the going away of the state of the Persians and the appearance of the Arab state ⟦9⟧ and the shift of dominion to them, and affairs will continue to be unstable and frightening ⟦10⟧ until the conjunction passed through the triplicity and returned to Scorpio in ⟦11⟧ the fourth conjunction. So the king of Persia went away where ⟦12⟧ there was no relation or connection between the two transits. But as for this adjoining shift, it is ⟦13⟧ as we described the preceding one. Verily, Māshā'allāh said that the lives of its kings will be lengthened ⟦14⟧ in the triplicity of Scorpio [*read* Sagittarius] because of the strong indication of the triplicities of Sagittarius concerning the king and ⟦15⟧ the kingdom. The state will become unstable during

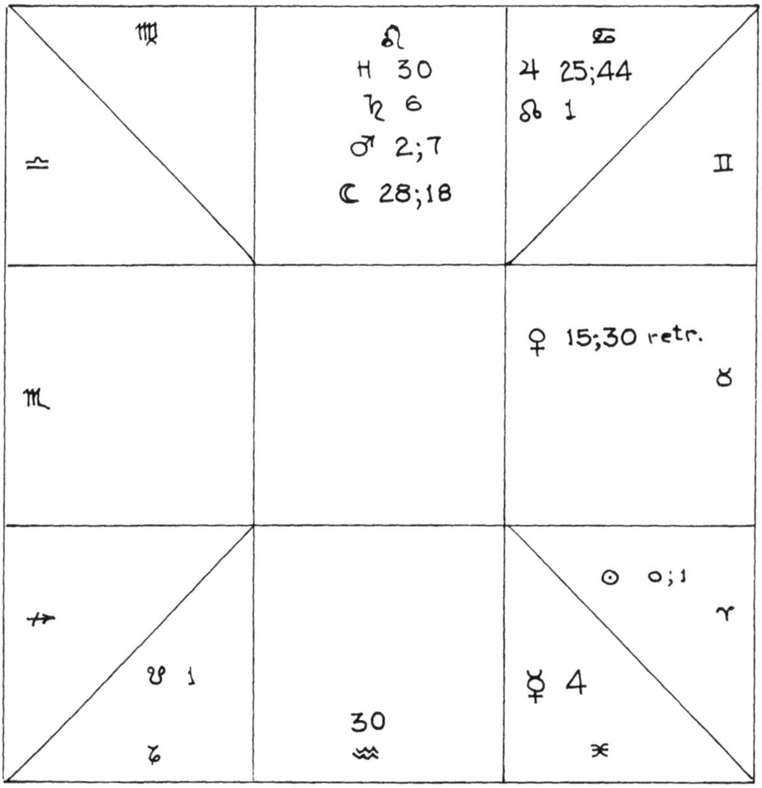

Horoscope 11.

the last years of the seventh conjunction when Saturn and Mars conjoin ⟦16⟧ in Cancer and the cycle returns to Sagittarius. Then a change in the king who ⟦17⟧ comes at this time, after fifteen years, is to be feared. He lifts up the sword at the end ⟦18⟧ of his career and rulership, against people. And in the eighteenth year of the seventh conjunction ⟦19⟧ the people will raid one another, and there will be wars until the eighth conjunction. ⟦f 219v, 1⟧ There passed after the first conjunction of this transit the second conjunction. The beginning of it was ⟦2⟧ in the year one hundred and ninety-seven of Yazdijird, and its duration was ⟦3⟧ nineteen years, ten months, and eleven days, solar, and the form of the heaven at ⟦4⟧ the entry of the sun into Aries for the year in which it occurred is according to this picture [*Horoscope 11*]. ⟦5⟧ Māshā'allāh said that when he looked at this picture, he found

⟦f 220r, 1⟧ the ascendant to be of Leo in its last degrees, and Mars and Saturn in the sign, ⟦2⟧ falling from the cardine, and the moon in the ascendant conferring counsel upon Mercury, and the sun ⟦3⟧ in the ninth conferring counsel upon Saturn, and Jupiter in its exaltation falling from ⟦4⟧ the twelfth house, and Mercury in the eighth entering the aspect, ⟦5⟧ and Venus, retrograding in upper midheaven, and conferring counsel upon Mars, which is ⟦6⟧ against it in heaven and house, she being not received, and Mercury was in ⟦7⟧ its cardine and in its dejection. So what he described indicates much killing and death in ⟦8⟧ the eastern region, and there appears in the first years of the conjunction an opponent for the sultan in these regions, and ⟦9⟧ he will interfere with him and will dominate the first sultan, and will call people to a religion other than ⟦10⟧ their religion, and they will degenerate morally, and their faith will disappear, and they will seek vice, and ⟦11⟧ the subjects (of the king) will become greedy and respect for the sultan and kings will disappear from their hearts, and ⟦12⟧ the wicked will become strong and the virtuous weak, most of whom will die, and women will elevate themselves ⟦13⟧ over the men, and vice will become prevalent, and mercy will disappear, and evil will come ⟦14⟧ from the East, and will be strengthened by it. The appearance of him who brings it will be at the beginning of ⟦15⟧ this year. He will pretend falsely to prophesy, and the consequence of his affair is corruption. ⟦16⟧ All that will be in the first year of the conjunction. And in the second ⟦17⟧ will be fighting and bloodshed. In the third will be death – annihilation by pestilence, ⟦18⟧ and spoiling of agriculture. And in the eighth year will also be killing, ⟦19⟧ and in the ninth increase of locusts and spoiling of the crops and plants and the climates will be ⟦f 220v, 1⟧ spoiled, all of them. And the third conjunction (will be) in the year two-hundred and nineteen [*read* 216] ⟦2⟧ of Yazdijird, and its duration nineteen years, ten months, and eleven days. ⟦3⟧ This is the form of the heaven at the transfer of the world-year in which it ⟦4⟧ took place [*Horoscope 12*]. ⟦5⟧ Māshā'allāh explained that he found in the transfer of this conjunction no star more fit for the guardianship than ⟦6⟧ the sun, because it is the lord of the ascendant and the transfer is diurnal and it (the sun) is safe from the maleficent (planets) ⟦7⟧ indicating the well-being of the fourth climate and its long duration and the goodness of its situation. ⟦f 221r, 1⟧ There

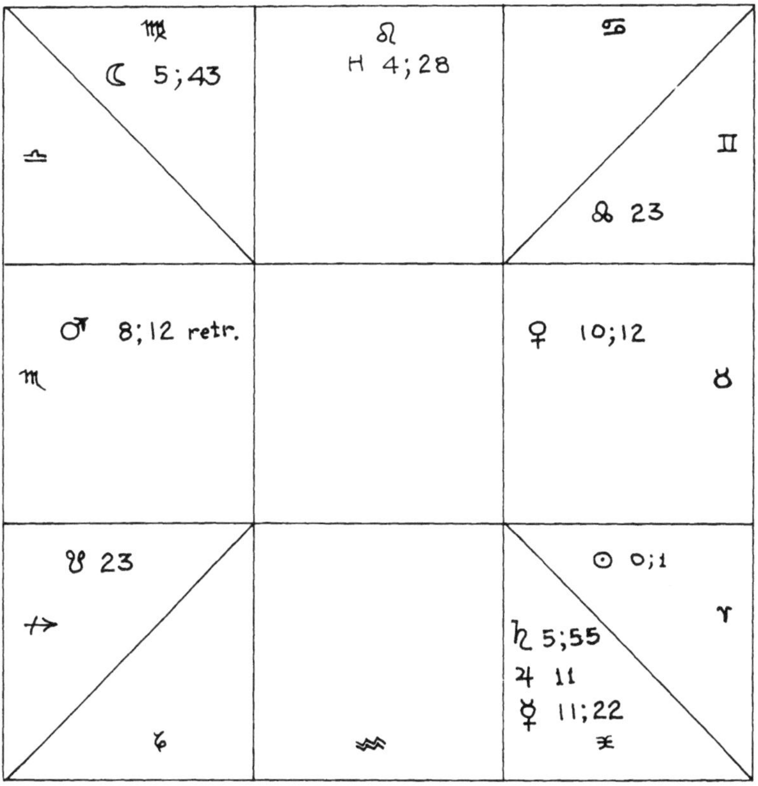

Horoscope 12.

will be in the eastern region death and war, and many will die in the cities ⟦2⟧ from a famine and the spoiling of their fruits in the second and the third and fourth years of ⟦3⟧ the conjunction. In the first and fifth there will be death and war among the Arabs. ⟦4⟧ In the first and second there will be locusts, and the cold will spoil the plants, and ⟦5⟧ death will be widespread in the land of Persia and Andalusia. And in the eighth year the people of ⟦6⟧ Ṭabaristān will encounter violence; then they will be victorious over their enemies, and their land will become fertile, and their cattle will improve, and ⟦7⟧ so God, be He exalted and glorified, is the most knowing and the best Judge. Then will be the fourth conjunction, and its duration ⟦8⟧ (is) nineteen years, ten months, and eleven days, and the first of it is in the year ⟦9⟧ two hundred thirty-seven [*read* 236] of Yazdijird. The form of the heaven at the transfer of the ⟦10⟧ world-year in which it took place is ⟦11⟧ what follows with this picture, if ⟦12⟧ God wills, be He exalted and glorified.

⟦f 221v⟧ [*Horoscope 13*] ⟦1⟧ Māshā'allāh explained that the conjunction occurred in Sagittarius, the cardines of the ascendant are perpendicular, and the planets are ⟦2⟧ cadent, except the sun. It is in the seventh, between it and the center of the seventh ⟦3⟧ there being two degrees, conferring counsel upon Saturn. Mars is in the ninth, and the moon is in the ⟦4⟧ fifth conferring counsel upon Jupiter, but it is not received. So that indicates that ⟦5⟧ the king of the fourth climate will be killed in the fourth year of this conjunction, ⟦6⟧ and the kingship will shift from the people of his house, and there will be at Isfahān and its regions a violent butchery, ⟦7⟧ and the king of the fourth region will travel, and there will occur a

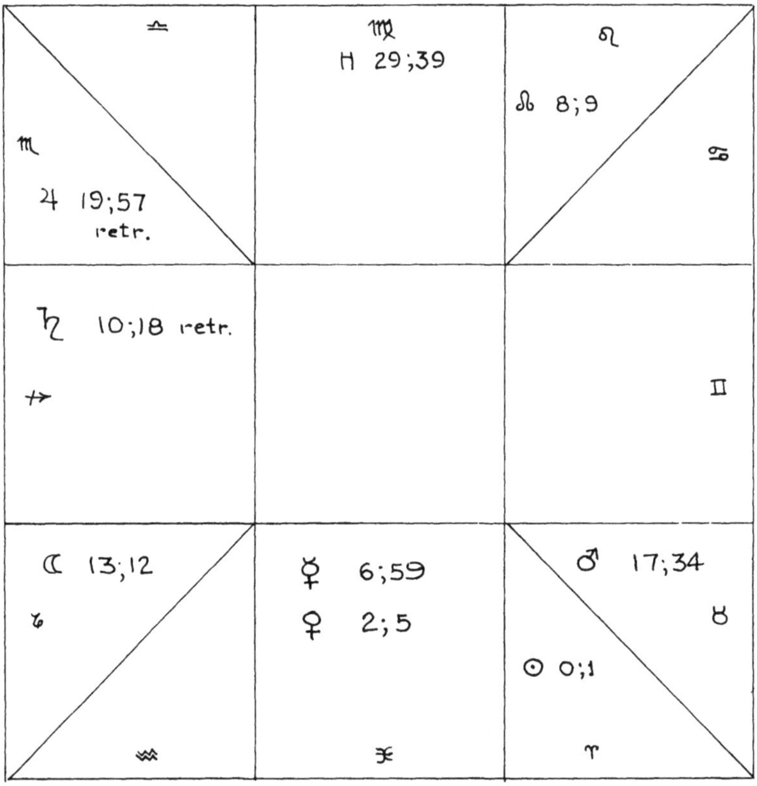

Horoscope 13.

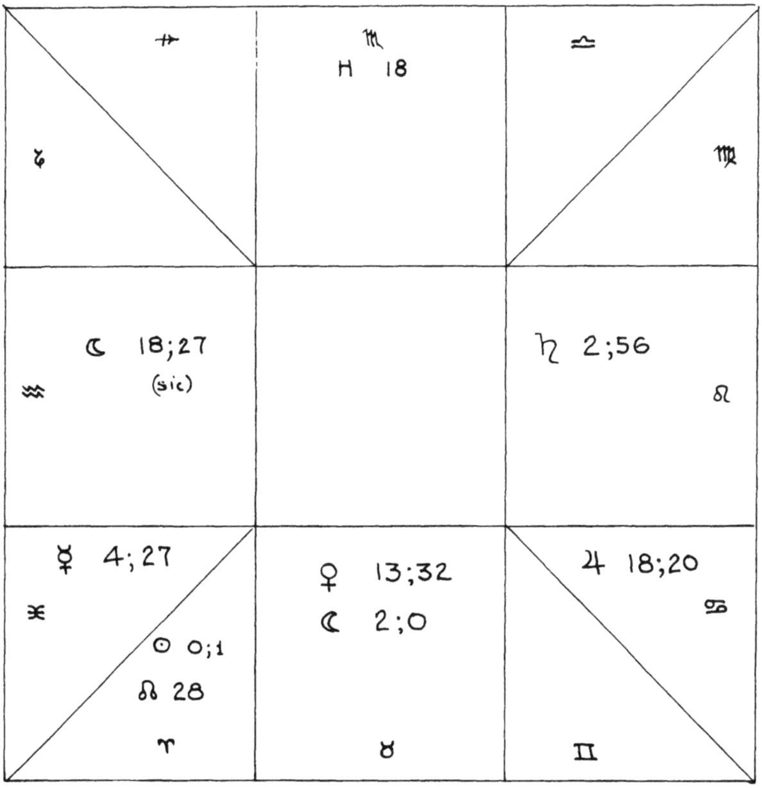

Horoscope 14.

dispute in the land of the Arabs, ⟦f 222r, 1⟧ and there will be degradation and shedding of blood. There will be in the sixteenth year a ⟦2⟧ mighty (and) general pestilence, and there will be battles in the second climate and dispute ⟦3⟧ with regard to religion, and the shedding of much blood therein, and victory and success ⟦4⟧ for the nobility after violence and misfortune in the fourth year. Most of the death ⟦5⟧ and pestilence will be in the sixteenth year in the regions of the West to the extent that it reaches ⟦6⟧ the regions of Mawṣil (Mosul). Several kings will die during this conjunction, and their situations will be straitened, ⟦7⟧ and animals will die in the first year of it; most of that (will be) among crawling animals ⟦8⟧ and cattle, but goats will be safe. The first climate will be safe, and (also) the seventh climate, ⟦9⟧ and there will occur in the second climate a dispute involving its

people. Then they will triumph. And in ⟦10⟧ the fifth and sixth (climates) there will be death. Of the climates (the one) in the worst condition will be the third because ⟦11⟧ misfortune enters it. It indicates that the obedience of the people of the fourth climate will improve. ⟦12⟧ The fifth conjunction will be in the two hundred and fifty-seventh year of Yazdijird, and its duration will be ⟦13⟧ nineteen years, ten months, and eleven days. ⟦14⟧ The form of the heaven at the entry of the sun into Aries in that year in which it occurred (is this):
⟦f 222v⟧ [*Horoscope 14*]

⟦1⟧ Māshā'allāh explained that when he looked at the year-transfer of this conjunction he found ⟦2⟧ of the planets Mars as having most witnesses in it and Venus strongest as to position; she ⟦3⟧ confers her strength upon Jupiter, and it (is) not received, and the moon, the light of the night, ⟦4⟧ confers its strength upon Saturn, which receives it from it (the moon). The place of the sun and the moon ⟦5⟧ with respect to Saturn indicates hardships which strike kings in this conjunction, such as ⟦6⟧ sickness and wars and a multiplicity of enemies, and (also) mighty war and killing ⟦f 223r, 1⟧ in the region of Khurāsān. Then it will become general among the people and especially in ⟦2⟧ the seventh climate. The enemy will appear in the fourth climate, and the rabble will set itself ⟦3⟧ above the kings. These wars and events and violence will be in ⟦4⟧ the tenth year, and what strikes the kings of the fourth climate will be in the fourth year. ⟦5⟧ The situation of the people of the fifth climate and the sixth climate and the second will improve, and ⟦6⟧ they will be blessed with peace and security from wars, and the situation of their leaders and ⟦7⟧ the influential people among them will be improved, except that the tenth year will be difficult and straitened, ⟦8⟧ and its evil will be general among people because it makes the rabble and the lower classes go against the nobility ⟦9⟧ and the kings during it. The sixth conjunction will be after the passing of ⟦10⟧ seventeen days of the third [*read* second] month of the year two hundred and seventy-seven of Yazdijird, and the duration (is) nineteen years, ten months, and eleven days. ⟦12⟧ The form of the heaven at the transfer of the world-year in which it took place ⟦13⟧ is thus: ⟦f 223v⟧ [*Horoscope 15*]

⟦1⟧ One of the calculators, he being 'Abdallāh b. Muḥammad b. Bishr the Astrologer, explained that ⟦2⟧ he examined Venus

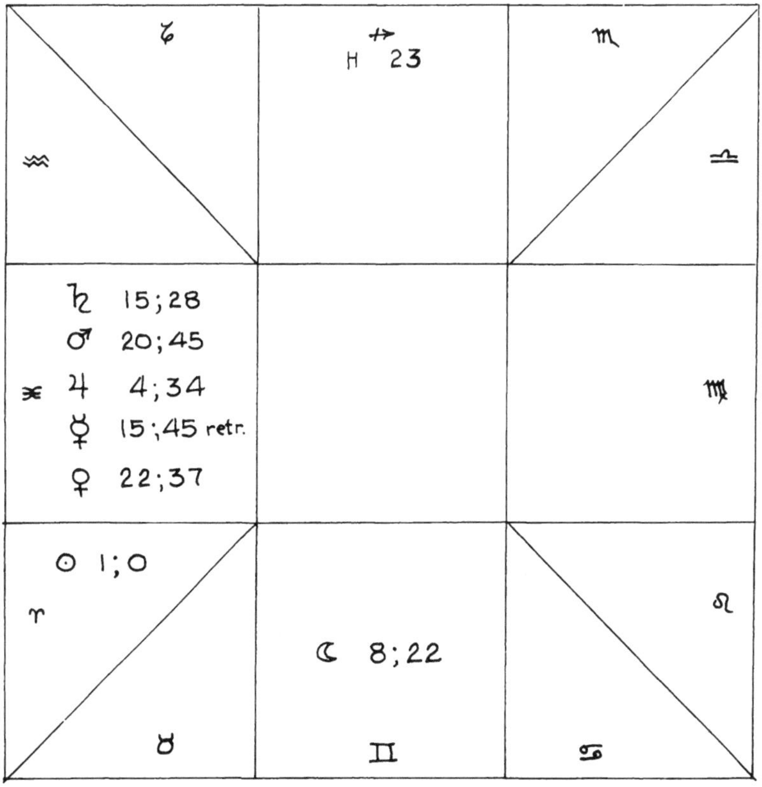

Horoscope 15.

and found it (to be) in Taurus 5;22,30, and that Māshā'allāh ⟦3⟧ fixed (?) it, in the chapter on it, because he should have added its equation, it being 30;22,30 to the ⟦4⟧ position of the sun, but he subtracted it from it. And (furthermore) that he examined the ascendant by adding to the ascendant of ⟦5⟧ the past conjunction what Māshā'allāh was adding of degrees for each year, it being ⟦6⟧ ninety-three degrees and fifteen minutes, for nineteen years. ⟦7⟧ So the ascendant of this conjunction came out as Libra 25. Our objective ⟦f 224r, 1⟧ in what we explained in this comment is to show how the ascendants of the conjunctions are obtained ⟦2⟧ according to the doctrine of Māshā'allāh concerning them and what it is that is added to the ascendant of ⟦3⟧ the past conjunction for each year so that there comes out the ascendant of the following conjunction. This ⟦4⟧ computation (is according) to the (Zīj al-)Shāh, because it is the zīj of Māshā'allāh, with which he was operating. ⟦5⟧ Māshā'allāh said that he looked at the transfer of this conjunction and he found the five ⟦6⟧ planets in the fourth, it being Pisces, and the moon in the seventh cadent, conferring ⟦7⟧ its strength upon Saturn. The sun in the fifth will be at the division in the fourth, and it is ⟦8⟧ cadent, not received, and it [*read* Venus] and Mars are under the rays, and Jupiter is eastern in ⟦9⟧ its house, and it is the lord of the ascendant of the conjunction, which indicates that the people will encounter ⟦10⟧ violence to the extent that it affects aquatic animals, causing blindness and ⟦11⟧ misfortunes among them. Evil will be general among people, and they will separate, one from another, and the beginning of ⟦12⟧ that will be in the region of the south (*qibla*), and it will spread around there, then it will take the direction of ⟦13⟧ the western region, and will spread on the earth (?*arḍayn*), and the people will encounter drowning because of much ⟦14⟧ water, and violence corrupts all the climates and the beginning of the spoliation will be in ⟦15⟧ the fifth and third climates, and kings will be destroyed in the fourth year of this ⟦16⟧ conjunction, and there will be in these years disputes and much shedding of blood, ⟦17⟧ until not one of the men of war will remain. And the people of the house of ⟦18⟧ prophecy and science will be strengthened, and good will be apparent from evil as white is distinguished from ⟦19⟧ black, and there will be sound growth among the people, and there will die, of the people in ⟦f 224v, 1⟧ the Western region in the seventh and sixteenth years, (so many) that ⟦2⟧ not one will remain long. There will remain with the kings no soldiers, and people will be scarce, and they will congregate ⟦3⟧ together. The seventh conjunction (will be) on the first of the first month ⟦4⟧ of the year two hundred and seventy-nine [*should be* 297] of Yazdijird. The form of the heaven ⟦5⟧ at the transfer of the world-year in which it occurs is thus [*Horoscope 16*].

⟦f 225r, 1⟧ Māshā'allāh said that he looked at the transfer of this year, and he found ⟦2⟧ the sun lord of the turn (light? *nawba*] in upper midheaven; it will be by division in the eleventh, and ⟦3⟧ it is conferring counsel upon Jupiter, and Jupiter returns the light ⟦4⟧ unto Saturn, in the sixth position, and it is received. The moon is lord of ⟦5⟧ the ascendant, cadent from it in the eighth, void of motion, conferring its

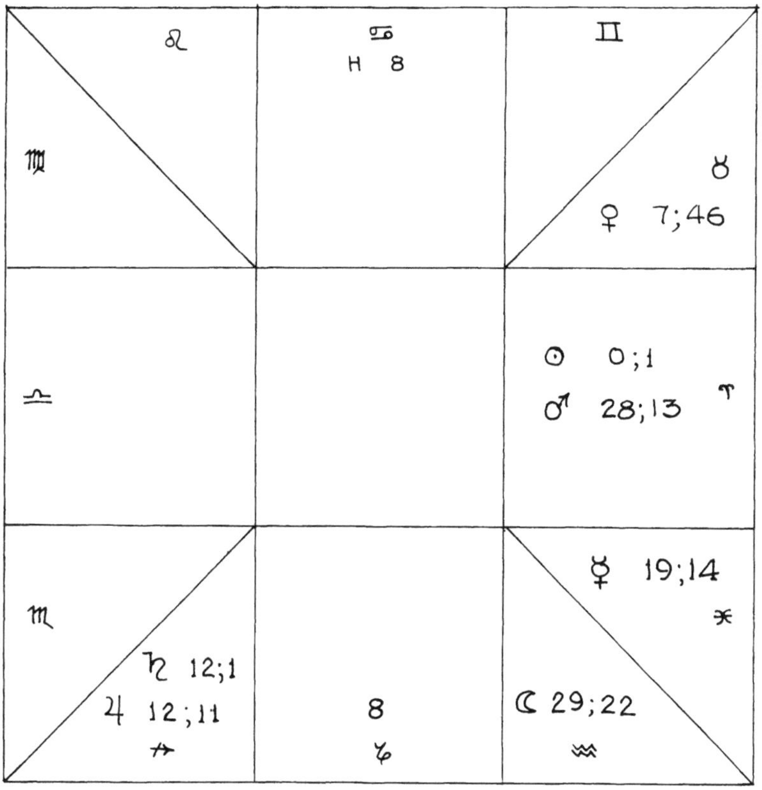

Horoscope 16.

counsel in ⟦6⟧ the second sign upon Jupiter, and it receives it. And Mars is in the tenth. Venus is ⟦7⟧ in the eleventh in its house, secure from the maleficent (planets). Mercury is in its descent ⟦8⟧ in the ninth, by equalization in the tenth. So all that indicates sicknesses ⟦9⟧ (which) will be in the climate of Babylon, and sorrow will enter upon its king in its sixth year. ⟦10⟧ Because of the place of Venus and the moon it indicates the security of the subjects and the goodness of their situation. ⟦11⟧ The place of Mercury indicates fame and high status obtained by the people of science and clerks, ⟦12⟧ and those who survived the first conjunction. Undoubtedly there will ⟦13⟧ be sickness among the people also in this conjunction, and eczema in the sixth year, and the eighteenth. ⟦14⟧ Verily this judgment is doubtful, because the other (things) he said about the sixth conjunction ⟦15⟧ we saw in the seventh conjunction, and I think that the copyist of Māshā'allāh's books interchanged ⟦16⟧ the two judgments so that he made the seventh judgment the sixth, or Māshā'allāh did ⟦17⟧ that on purpose, for he frequently misleads (people). However, we ⟦18⟧ have put it as we found it. And we say after that as to what he witnessed concerning the situation of ⟦19⟧ the seventh conjunction, that the first of the world-year in which it took place occurred on the first of ⟦f 225v, 1⟧ year two hundred and ninety-seven of Yazdijird, and the time of the conjunction of ⟦2⟧ the two planets by the Mumtaḥan (Zīj) was at Sagittarius 16;41 after six months and nine days ⟦3⟧ from the year-transfer, and Mars was at the time in ⟦4⟧ Virgo 16;25, and it was in the tenth from the two (planets). The sun (was) in Libra 21;21. ⟦5⟧ So Mars is elevated above the two and in its sector ascending from the orb of its apogee, and from ⟦6⟧ the orb of its epicycle east, ascending in its latitute to the north also. Saturn is passing in ⟦7⟧ the orb of its apogee over Jupiter, and since (this) was on the first of the eleventh month of this ⟦8⟧ year, the soldiers revolted (?) against (the caliph) al-Muqtadir b'Illāh, and they deposed him. So Nāzūk was killed ⟦9⟧ and Abū al-Hayjā (also), and that crowd was dispersed and al-Muqtadir returned to his place. After ⟦10⟧ that by one year the men killed in their dwelling places (?) at one time a few ⟦11⟧ thousand people, and strife and rebellion continued until the eighth month of ⟦12⟧ the fifth year of the conjunction. There was killed in it al-Muqtadir b'Illāh, and al-Qāhir replaced him. ⟦13⟧ And when it was the second month of the seventh year of the conjunction, al-Qāhir ⟦14⟧ was deposed, and his eyes were put out and al-Rāḍī succeeded to the caliphate in his place. ⟦15⟧ In his days were wars and strife and killing of the chiefs of the Ḥajariya and the Sājiya (units of the army). ⟦16⟧ The remaining (ones) dispersed, and they were about four thousand young men, and the vizirate ceased. ⟦17⟧ The men of the sword seized control of affairs until in the first of ⟦18⟧ the tenth month of the thirteenth year of it there died al-Rāḍī b'Illāh, ⟦19⟧ and al-Muttaqī succeeded him in his place. There occurred in his days strife and catastrophes until the end of the seventh month. (*In the MS as bound, the misplaced folios follow this.*)

3. ON THE CHRONOLOGY AND ASTRONOMY OF MĀSHĀ'ALLĀH'S ASTROLOGICAL HISTORY

The framework into which Māshā'allāh attempts to fit world history is a series of conjunctions of the two planets farthest from the Earth in ancient and medieval astronomy – Saturn and Jupiter. In doing this he is clearly following a Sasanian source.[1] Since Saturn travels at a rate of about 12° a year and Jupiter about 30°, a mean conjunction of the two planets must take place at a point on the zodiac having a longitude of about 240° more than that at which the previous mean conjunction occurred and about 20 years later; for

$$12^{\circ/y} \cdot 20^{y} = 240^{\circ}$$

$$30^{\circ/y} \cdot 20^{y} = 600^{\circ} = 360^{\circ} + 240^{\circ}.$$

In fact, the parameters which Māshā'allāh mentions are $19^{y}\ 10^{m}\ 11^{d}$ (f 215r:4-5 *et al.*; he sometimes shortens this by one day) and ⟨24⟩2;25° (f 215r:2-3). If his computations were accurate and if we knew the length of his year, we could easily derive the mean motions of Saturn and Jupiter from this information; but before proceeding further we must investigate the manner of computation.

One begins with mean conjunction 0, which occurs in ♉ 7;42° when $509^{y}\ 2^{m}\ 24^{d}$ of the 1000^{y} of Mars have passed (ff 214v:17-215r:1). At regular intervals of longitude and time after this other mean conjunctions occur. The ones which Ibn Hibintā reports with their longitudes and their times from conjunction 0 are the following:

1. See D. Pingree, "Historical Horoscopes," *JAOS* 82 (1962), 229-246.

Mean conjunction	*Longitude*	*Time*
0	♉ 7;42	0
121 (f 215r:5-9)	♏ 1;24	$2412^y\ 6^m\ 26^d$
122 (f 226r:8-9)	♋ 3;40	$2423^y\ 6^m\ 12^d$ (f 227v:5-9)
289 (f 227v:12-15)	♐ 18;55	$5749^y\ 5^m\ 8^d$
290 (f 228v:1-3)	♌ 11;20	
320 (ff 229v:14-230r:2)	♏ 4;2	$6336^y\ 0^m\ 26^d$
323 (ff 231r:19-231v:1)	♏ 11;19	
324 (f 232r:15-17)	♋ 13;42	

Computing longitudes with Māshā'allāh's parameter of ⟨24⟩2;25° per mean conjunction will immediately show that the true parameter is somewhat larger; but the divergent trends which result from using 242;25,35° demonstrate that Māshā'allāh is simply too sloppy a computer to be trusted at any point.

Mean conjunction	*Longitude*				
	Text	242;25°	Δ	242;25,35°	Δ
0	♉ 7;42°	♉ 7;42°	0°	♉ 7;42°	0°
121	♏ 1;24°	♏ 0;7°	−1;17°	♏ 1;17,35°	−0;6,25°
122	♋ 3;4⟨9⟩°	♋ 2;32°	−1;17°	♋ 3;43,10°	−0;5,50°
289	♐ 18;55°	♐ 16;7°	−2;48°	♐ 18;55,35°	+0;0,35°
290	♌ ⟨2⟩1;20°	♌ 18;32°	−2;48°	♌ 21;21,10°	+0;1,10°
320	♏ 4;2°	♏ 1;2°	−3°	♏ 4;8,40°	+0;6,40°
323	♏ 11;19°	♏ 8;17°	−3;2°	♏ 11;25,25°	+0;6,25°
324	♋ 13;4⟨4⟩°	♋ 10;42°	−3;2°	♋ 13;51,0°	+0;7°

When we compute the times at which the mean conjunctions occurred with Māshā'allāh's parameter of $19^y\ 10^m\ 11^d$ per mean conjunction we find even greater confusion. In computing the following table it is assumed that $30^d = 1^m$ and $12^m = 1^y$.

Mean conjunction	*Time*	
	Text	$19^y\ 10^m\ 11^d$
0	0	0

121	$2412^y\ 6^m\ 26^d$	$2403^y\ 6^m\ 11^d$
122	[$2423^y\ 6^m\ 12^d$]	$2423^y\ 4^m\ 22^d$
289	$5749^y\ 5^m\ 8^d$	$5740^y\ 7^m\ 29^d$

What one immediately suspects is that the interval of $9^y\ 2^m\ 24^d$ between 500^y of the 1000^y of Mars and mean conjunction 0 is involved in the text entries for conjunctions 121 and 289 (cf. f 227v: 5-9 and f 229v: 8-10). Subtracting this amount, one gets respectively:

121	$2403^y\ 4^m\ 2^d$
289	$5740^y\ 2^m\ 18^d$

If one assumes that the parameter used by Māshā'allāh was $19^y\ 10^m\ 10^d\ 11^h$ or $19^y\ 10^m\ 10^d\ 10^h$, one finds for these conjunctions nearly perfect agreement:

	$19^y\ 10^m\ 10^d\ 11^h$	$19^y\ 10^m\ 10^d\ 10^h$
121	$2403^y\ 4^m\ 5^d$	$2403^y\ 4^m\ 0^d$
289	$5740^y\ 2^m\ 22^d$	$5740^y\ 2^m\ 10^d$

Now 122 conjunctions times $19^y\ 10^m\ 10^d\ 11^h$ would give $2423^y\ 2^m\ 16^d$. In fact, we know from the text that the figure given there – $2423^y\ 6^m\ 12^d$ – was computed by Ibn Hibintā, presumably using the rougher parameter $19^y\ 10^m\ 11^d$. Ibn Hibintā has made an error in his calculation.

This is not his only error. Conjunction 290 precedes the year in which Christ was born by 13^y, and we are informed by the text that the interval between the midpoint of the 1000^y of Mars and the year in which Christ was born is 5773^y ⟨2⟩4^d (f229v:6-10). If we multiply $19^y\ 10^m$ by 290 (ignoring the days), we get $5751^y\ 8^m$; to this must be added $9^y\ 2^m\ 24^d$ (from the midpoint of the 1000^y of Mars to conjunction 0) and 13^y. This sum equals $5773^y\ 10^m\ 24^d$. Ibn Hibintā has simply dropped the 10^m.

This is proved by his further statement that the interval between the Deluge (conjunction 122) and the year in which Christ was born is $3349^y\ 6^m\ 12^d$ (f 229v:11-13). This number he derives by subtracting $2423^y\ 6^m\ 12^d$ – his erroneous interval between conjunctions 0 and 122 – from $5773^y\ 0^m\ 24^d$. In this also he ignores the fact that this latter number of years is supposed to represent the interval between the midpoint of the 1000^y of Mars and the year in which Christ was born, not between conjunction 0 and that last date.

However, the series of errors does not end here. We are next told (apparently on the authority of Māshā'allāh) that there are $6345^y\ 3^m\ 20^d$ between the midpoint of the 1000^y of Mars and the mean conjunction which indicated the birth of Muḥammad[2] (f 229v:15-16). It is known that the conjunction occurred in A.D. 571; and, if we subtract $5773^y\ 0^m\ 24^d$ from $6345^y\ 3^m\ 20^d$, we get $572^y\ 2^m\ 26^d$ as the interval between the beginning of the year in which Christ was born and the conjunction of 571. In other words, the birth of Christ is correctly put in -1. However, it must be remembered that the horoscope of the vernal equinox of the year in which Christ was born is dated -12. So here Māshā'allāh has clumsily jammed together two chronological systems – one of which dates the birth of Christ in -12, the other in -1 – without any regard for their fundamental incongruence.

The number $6345^y\ 3^m\ 20^d$ is, of course, $(19^y\ 10^m\ 10^d\ 11^h) \cdot x - (9^y\ 2^m\ 24^d)$. Therefore, to find the interval between conjunction 0 and conjunction x Ibn Hibintā correctly subtracts, and gets $6336^y\ 0^m\ 24^d$. Now in fact $319 \cdot (19^y\ 10^m\ 10^d\ 11^h) = 6336^y\ 1^m\ 6^d$ and $319 \cdot (19^y\ 10^m\ 10^d) = 6336^y\ 0^m\ 22^d$. But in fact the conjunction of 571 is conjunction 320, and the correct figures should be:

$$320 \cdot (19^y\ 10^m\ 10^d\ 11^h) = 6355^y\ 11^m\ 17^d$$

or

$$320 \cdot (19^y\ 10^m\ 10^d\ 10^h) = 6355^y\ 11^m\ 3^d.$$

Thus, in summary of this part of our investigation, we can say that Māshā'allāh's chronology depends on a millennial theory rather than on a conjunction theory, for the intervals he gives are all from the midpoint of the millennium of Mars to a particular event. In this he differs from other astrologers, such as Abū Ma'shar in his *Kitāb al-ulūf*, who begin their histories with the Deluge (the beginning of Kaliyuga) marked by the mean conjunction of all the planets in ♈ 0° (Indian) at about the vernal equinox of -3101. Thus, Māshā'allāh's basic system may be illustrated by Fig. 1.

2. It may be noted here that there is a tradition, ostensibly of the first century of the Hijra, that, of the world's 7000 years, 6000 and some hundreds had already passed when Muḥammad was born; see *L'Abrégé des merveilles*, trans. Carra de Vaux (Paris, 1898), pp. 5-6.

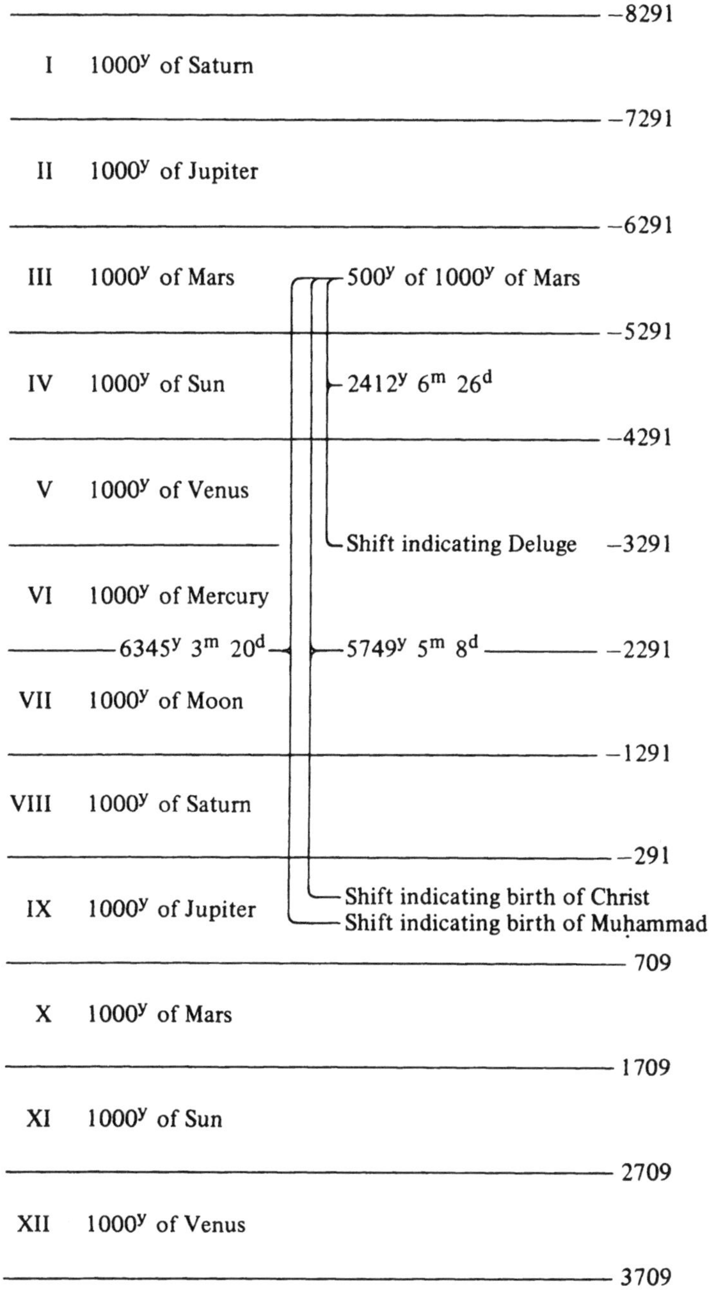

Fig. 1. Māshā'allāh's Chronology.

From this chart the influence of Zoroastrian ideas will be immediately apparent, though they are interpreted to conform with the necessities of the astrological theory of Jupiter–Saturn conjunctions and to confirm the Islamic rather than the Zurvanite or Mazdean revelation. Some Zoroastrians, such as the author(s) of the first chapter of the Bundahishn, divide the 12000^y of the creation into four equal periods; during the first three millennia – before the assault of Ahriman and the daevas – there is no motion in the heavens; between the fourth and ninth millennia there is a famous struggle between the forces of Light and Darkness, with some sort of catastrophic event in the sixth millennium; and in the last three millennia the Good finally prevails.[3] Now, if one shifts Māshā'allāh's scale down 1000^y – that is, if one places its beginning in the middle of the Sun's 1000^y rather than in that of Mars – one finds perfect correspondence between the Zoroastrian doctrine and Māshā'allāh. The motion of the heavens commences in the fourth millennium (after 3509^y), the Deluge – a catastrophic event – occurs at the end of the sixth millennium (after 5932^y) and Christ and Muḥammad, who defeat evil, were born in the tenth millenium (after 9282^y and 9845^y, respectively). That the association of the Deluge with Cancer reflects Zoroastrian thought has been shown elsewhere.[4] It may seem rather drastic to displace conjunction 0 from the millennium of Mars to that of the Sun. However, it must be realized that this is theologically very satisfactory not only because of the positions within the 12000^y period that it bestows upon the four crucial events, but also because one thereby has the sun – the prime symbol of light and the King of the planets – presiding over the activization of the material creation. Māshā'allāh's error, I assume, is due to his thinking that the

3. See, for example, H. Corbin, "Le temps cyclique dans le Mazdéisme et dans l'Ismaélisme," *Eranos-Jahrbuch* 20 (1951), 149-217, esp. pp. 155-156. For the beginning of the motions of the planets after 3000^y see Manushčihr in Dātastāni Dēnik, translated by M. Molé, *Culte, mythe, et cosmologie dans l'Iran ancien* (Annales du Musée Guimet, Bibliothèque d'études 69, Paris, 1963), p. 402, with which compare Bīrūnī, *Chronology*, published in *Documenta Islamica Inedita*, ed. J. Fück (Akademie Verlag, Berlin, 1952), p. 84.

4. See D. Pingree, "Astronomy and Astrology in India and Iran," *Isis* 54 (1963), 229-246, esp. 244.

motions of the planets must begin in the third 1000^{y} rather than after 3000^{y}. He may also have been influenced by someone's desire to transfer the responsibility for Ahriman's assault from the Sun to the evil planet, Mars.

Onto this essentially Zoroastrian scheme of things Māshā'allāh has grafted the Sasanian theory of the possibility of writing history on the basis of Jupiter–Saturn conjunctions; and, despite a few errors in computation which we have noted above, he has succeeded in making the two systems approximately coincide. However, whereas all other conjunction-histories at present known (as has been remarked above) use as their initial point the epoch of the present Kaliyuga (17/18 February −3101) when there was a mean conjunction of all the planets in ♈ 0° of the (Indian) fixed zodiac, Māshā'allāh in order better to accommodate Zoroastrian doctrine, begins with a much earlier Jupiter–Saturn conjunction in ♉ 7;42° which supposedly occurred 9^{y} 2^{m} 24^{d} after 500^{y} of the 1000^{y} of Mars (*read* the Sun) has passed. We are told that the event was about 5764^{y} before the year in which Christ was born; but its precise dating must depend on our dating of the various horoscopes given by Māshā'allāh and on our determination of the length of his year.

As far as the latter problem is concerned, it is at once apparent from the horoscopes that the year is solar and not Egyptian-Persian, for they are computed for vernal equinoxes. It is only to be determined whether the solar year employed is tropical or sidereal. Ibn Hibintā's statement that Māshā'allāh used the *Zīj al-Shāh* and an annual excess of revolution equivalent to 93;15° would indicate that it is sidereal; for the *Zīj* employs and the parameter implies a year-length of $6,5;15,32,30^{d}$.

One can confirm this result by examining the dates of the last eleven horoscopes in the series (the exact day depends upon the position of the Moon) and subtracting the longitudes of the Sun according to Māshā'allāh – all, of course, are ♈ 0° – from the computed longitudes:

Conjunction	*Date*	*(Computed Sun) −(Text Sun)*	*Ascendant*
320	19 March 571	+1	♎ 22
323	19 March 630	+1	♑ 18
324	19 March 650	+2	♈ 10
329	19 March 749	+2	♏ 25

Conjunction	*Date*	*(Computed Sun) –(Text Sun)*	*Ascendant*
332	20 March 809	+4	♋ 3
333	20 March 829	+4	♌ 30
334	20 March 848	+4	♌ 4;28
335	20 March 868	+5	♍ 19;39
336	20 March 888	+5	♏ 18
337	21 March 907	+5	♐
338	21 March 928	+6	♋ 8

The number of Julian days between 19 March 571 and 21 March 928 is 130397^d; $6,5;15,32,30 \cdot 5,57 = 130397^d$.

And, as has been noted by 'Abdallāh ibn Muḥammad ibn Bishr, though his calculations were wrong, the same parameter should be detectable from the ascendants. In fact, this is sometimes the case; in the absence of secure knowledge of the rising-times employed by Māshā'allāh, the following ascendants have been computed with the useful Babylonian System A which was also used by the Indian sources of the *Zīj al-Shāh*:

♓ ♈ 20°
♒ ♉ 24°
♑ ♊ 28°
♐ ♋ 32°
♏ ♌ 36°
♎ ♍ 40°

(See the Commentary on f 221v: 1-3).

Conjunction and year	*Text*	*Ascendants*: *93° (with rising times)*		*93° (without rising times)*
320 (571)	♎ 22			
323 (630)	♑ 18	♎ 22+96;24	= ♑ 28;24	♎ 22+87 = ♑ 19
324 (650)	♈ 10	♑ 18+47;12	= ♓ 5;12	♑ 18+60 = ♓ 18
329 (749)	♏ 25	♈ 10+221;44	= ♏ 21;44	♈ 10+207 = ♏ 17
332 (809)	♋ 3	♏ 25+150	= ♈ 25	♏ 25+180 = ♉ 25
333 (829)	♌ 30	♋ 3+68;48	= ♍ 11;48	♋ 3+60 = ♍ 3

334 (848)	♌4;28	♌30+320;48	=♋20;48	♌30+327 =♋27
335 (868)	♍19;39	♌4;28+76;52	=♎20;20	♌4;28+60 = ♎4;28
336 (888)	♏18	♍19;39+77;23	= ♐7;2	♍19;39+60 =♏19;39
337 (907)	♐23	♏18+318;24	= ♎6;24	♏18+327 =♎15
338 (928)	♋8	♐23+126;24	=♈29;24	♐23+153 = ♉26

From these figures it should be clear that Māshā'allāh has computed the longitudes of the ascendants by using an annual excess of revolution close to 93°, but that he has ignored completely the effect of the obliquity of the ecliptic. In other words, his "ascendants" correspond to arcs along the equator (divided into signs, degrees, and minutes) rather than along the ecliptic, and thereby – if by nothing else – his whole procedure is vitiated. But even the equatorial arcs are not accurately computed.

Having established that the length of Māshā'allāh's year is $6,5;15,32,30^d$, we can now say that the interval between mean conjunctions is $(6,5;15,32,30 \cdot 19) + 5,10;25^d$ (using $10^m\ 10^d\ 10^h$) or $2,0,50;20,17,30^d = 7250^d\ 8^h+$. With this parameter we can easily compute the number of Julian years between the various mean conjunctions:

Conjunction	*Text years*	*Julian years*	*Date (A.D.)*
Half of 1000^y of Mars	$-9^y\ 2^m\ 24^d$	$-9^y\ 84^d$	2 Aug. –5791
0	0	0	3 Nov. –5782
121	$2403^y\ 4^m\ 2^d$	$2401^y\ 325^d$	25 Sept. –3380
122 (Deluge)	$2423^y\ 6^m\ 12^d$	$2421^y\ 271^d$	31 July –3360
289	$5740^y\ 2^m\ 18^d$	$5736^y\ 273^d$	1 Aug. –45
290		$5756^y\ 219^d$	9 June –25
290 + 13^y (Christ)		$5769^y\ 219^d$	VE –12
320	$6336^y\ 0^m\ 26^d$	$6352^y\ 40^d$	12 Dec. 570
337		$6689^y\ 206^d$	27 May 908 (so text)
338		$6709^y\ 152^d$	3 April 928 (so text)

These dates indicate that the Deluge occurred in the year –3360, or just 259^y (= 13 mean conjunctions) before the conjunction that marked the beginning of the present Kaliyuga on 17/18 February –3101. It is probable that the 249^y difference between the dates of the Deluge (–3101) according to

Abū Ma'shar, Kūshyār ibn Labbān, and the *Zīj al-Ma'mūnī* on the one hand and that according to the most precise chronologers on the other as recorded by Abū al-Fidā'[5] should really be 259^{y}. The date –3380 for the conjunction indicating the Deluge is also confirmed elsewhere by Abū Ma'shar (see *The Thousands of Abū Ma'shar*, pp. 41-42). Moreover, as mean conjunction 122 occurred in ♋3;49, it is easy to compute with our parameter $242;25,35^{\circ/conj}$ that mean conjunction 135 occurred in ♈5;21,35. And this proves that Māshā'allāh, though according to 'Abdallāh ibn Muḥammad ibn Bishr he used the *Zīj al-Shāh*, did not assume a mean conjunction of all the planets at ♈0° at the vernal equinox of –3101; therefore, neither did his version of the *Zīj al-Shāh*!

Having established this much, it is necessary to date the horoscopes before 571. This can now be easily done since they are all of vernal equinoxes of known years, and we also know the length of the year employed. The results of the calculations are as follows:

Conjunction	*Date of vernal equinox (A.D.)*	*Date of mean conjunction*
0		3 Nov. –5782
121	11 Feb. –3380	25 Sept. –3380
122	12 Feb. –3360	31 July –3360
289	13 Mar. –45	1 Aug. –45
290	14 Mar. –25	9 June –25
$290 + 13^{y}$	14 Mar. –12	
320	19 Mar. 571	12 Dec. 570
323	19 Mar. 630	
324	19 Mar. 650	
329	19 Mar. 749	
332	20 Mar. 809	
333	20 Mar. 829	
334	20 Mar. 848	
335	20 Mar. 868	
336	20 Mar. 888	
337	21 Mar. 907	27 May 908
338	21 Mar. 928	3 Apr. 928

It is now our task to derive the astronomical parameters upon which the computation of these horoscopes is based.

5. *Abul-Fedae Historia Anteislamica*, ed. H. O. Fleischer (Lipsiae, 1831), p. 4.

It is, of course, trivial to get mean motions of Saturn and Jupiter; one must simply divide 4,2;25,35° and 10,2;25,35°, respectively, by 2,0,50;20,17,30^d. The results are:

Saturn	0;2,0,22,17$^{\circ/d}$,
Jupiter	0;4,59,7,18$^{\circ/d}$.

Using these two parameters, it requires only a bit of work to construct the following table showing horoscope longitudes, mean longitudes, and equations of Saturn (one knows that on 3 November −5782 it was in ♉7;42).

Saturn

Date	*Horoscope longitude*	*Mean longitude*	*Equation*
11 Feb. −3380	♏1;56 retr.	♎23;43	+8;13
12 Feb. −3360	♊26;0	♊28;0	−2;0
13 Mar. −45	♐9;9	♐14;1	−4;52(!)
14 Mar. −25	⟨♌⟩	♌18;18	
14 Mar. −12	♑16;28	♑27;6	−10;38(!)
19 Mar. 571	♏6;13 retr.	♏7;14	−1;1
19 Mar. 609	♒9	♒21;18	−12;18
19 Mar. 630	♏4;30	♏7;45	−3;15
19 Mar. 650	♋2;2	♋12;2	−10;0
19 Mar. 749	♏19 retr.	♏21;4	−2;4
20 Mar. 809	♐	♐3;52	
20 Mar. 829	♌6	♌8;7	−2;7
20 Mar. 848	♓5;15	♈0;12	−24;57(!)
20 Mar. 868	♐10;18	♐4;27	+5;51(!)
20 Mar. 888	♌2;16	♌8;42	−6;26
21 Mar. 907	♓15;28	♈0;45	−15;17
21 Mar. 928	♐12;1	♐17;15	−5;14

We can correct some of Māshā'allāh's figures from a general knowledge of Saturn's mean motion without assuming anything about his specific value. In 59 years it makes only slightly more than two revolutions; in one year it progresses about 12°. Therefore the entry for 571 must be changed to

♏4;13; that for 829 to ♌⟨0⟩;6; that for 848 to ♓⟨1⟩5;15; and that for 868 to ♐0;18.

Jupiter

Date	*Horoscope longitude*	*Mean longitude*	*Equation*
11 Feb. –3380	♎12;39 retr.	♎12;33	+0;6
12 Feb. –3360	♊14;36	♊19;36	–5;0
13 Mar. –45	♏28;58	♐6;56	–7;58
14 Mar. –25	♋25;24	♌13;59	–18;35
14 Mar. –12	♍2;30 retr.	♍18;34	–16;4
19 Mar. 571	♏4;26	♏12;2	–7;36
19 Mar. 609	♑21	♑25;14	–4;14
19 Mar. 630	♎23;4	♏2;31	–9;27
19 Mar. 650	♊23;7	♋9;34	–16;27
19 Mar. 749	♏6;8 retr.	♏14;0	–7;54
20 Mar. 809	♏29;32	♐8;59	–9;28
20 Mar. 829	♋25;44	♌11;57	–16;13
20 Mar. 848	♓11	♓18;40	–7;40
20 Mar. 868	♏19;57	♏25;39	–5;42
20 Mar. 888	♋19;20	♌2;37	–14;17
21 Mar. 907	♓4;34	♓9;15	–4;14
21 Mar. 928	♐12;11	♐16;38	–3;47

From these data it is apparent that Māshā'allāh's mean longitudes of both Saturn and Jupiter used for computing their true longitudes in all the horoscopes after the second were about 10° lower than those given in the mean longitude columns computed above. This systematic deviation can be explained in the following manner. Our astrologer made an error of 10° in computing the longitude of mean conjunction 0; since the longitudes of all his mean conjunctions are computed from this initial erroneous value, they also are uniformly off by 10°. He also computed the longitudes of Saturn and Jupiter for the horoscopes of the vernal equinoxes of –3380 and –3360 from their mean position on 3 November –5782; but the remaining longitudes of these two planets are computed from the correct epoch longitudes of his *Zīj*. We shall investigate this matter further later.

For now, the only other parameter we can derive from the horoscopes is that of the motion of the lunar node:

Date	*Difference*	λ	Δλ	*Mean yearly motion*
−3380		♋ 8		
−3360	20	♊ 11	−27°	−19;21°
571	3951	♏ 7;53	−240;7°	−19;18,59,...°
630	4010	♓ 1	−127°	−19;20,0,...°
809	4189	♋ 28;55	−339;5°	−19;19,52,...°
829	4209	♋ 1	−7°	−19;19,45,...°
848	4228	♊ 23	−15°	−19;19,54,...°
868	4248	♌ 8;9	−329;51°	−19;18,53,...°
888	4268	♈ 28	−70	−19;19,56,...°

From this it appears that the mean yearly motion of the node is about −19;19,54°, and that the entries for the years 571, 630, and 868 are not correct (one need add only 0;22° to the entry for −3360 to correct it).

Date	*Text*	*Computation*	*Text − computation*
−3380	♋ 8		
−3360	♊ 11	♊ 11;22,0°	−0;22,0°
571	♏ 7;53	♉ 10;46,48	
630	♓ 1	♓ 8;1,0	−7;1,0°
809	♋ 28;55	♋ 27;38,54	+1;16,6°
829	♋ 1	♋ 1;0,54	−0;0,54°
848	♊ 23	♊ 23;42,48	−0;42,48°
868	♌ 8;9	♉ 27;4,48	
888	♈ 28	♉ 0;26,48	−2;26,48

It has been calculated from Bīrūnī's *Tamhīd al-mustaqirr lima'nā al-mamarr*[6] that Māshā'allāh's version of the *Zīj al-Shāh* gave the following maximum equations:

6. Page 87 of the Arabic text, p. 100 of the translation; see E. S. Kennedy, *JAOS* 78 (1958), 256.

Planet	Center	Anomaly
☉	2;13°	
☾	4;56°	
♄	8;37°	5;44°
♃	5;6°	10;52°
♂		40;31°
♀	2;13°	47;11°
☿	4;0°	21;30°

From the same book (99:5 - 103:14) we have a long discussion of Māshā'allāh's treatment of the horoscope of the year-transfer marking the shift of the conjunctions from the triplicity of earth to that of air (see the commentary on f 219r:5-13). This is mean conjunction 308; and the year-transfer referred to can be dated 18 March 333. Bīrūnī reports Māshā'allāh's positions only for the superior planets and the ascendant.

Planet	Text	Computation	Text – computation
♄	♎ 9;8	♎ 4	+5
♃	♍ 22;44	♎ 7	–14
♂	♓ 14	♒ 0	+44
☉	♈ 0;1	♓ 28	+2
H	♌ 20		

Furthermore, Bīrūnī in the same passage gives equations which he and Māshā'allāh found in their respective copies of the *Zīj al-Shāh* as well as the longitudes of the planets' apogees.

Planet	Equation of Center	Argument	Apogee
	Māshā'allāh		
♄	6;40°	50;52°	240°
♃	1;7°	12;44	160°
	2;28°	29;8°	
♂	9;26°	57;44°	115°
☉	2;5°	70;52°	80°

Planet	*Equation of Center*	*Argument*
	Bīrūnī	
♄	6;35°	50;52°
	6;41°	52°
♃	1;9°	12;44°
	2;28°	29;8°
♂	9;27°	57;44°
☉	2;6°	71°
	2;10°	80°

We are now in a position to determine the method by which the tables of the equations of the center in the *Zij al-Shāh* of Māshā'allāh and in that of Bīrūnī were computed. Values arrived at by the method of sines and by the method of declinations are compared:[7]

Māshā'allāh

Planet	*Argument*	*Method of sines*	*Method of declinations*	*Text*
Saturn (E_{max} = 8;37°)	50;52°	6;41,1°	6;36,10°	6;40°
Jupiter (E_{max} = 5;6°)	12;44°	1;7,27°	1;5,35°	1;7°
	29;8°	2;28,58°	2;25,37°	2;28°
Mars (E_{max} = 11;12°)	57;44°	9;28,12°	9;23,15°	9;26°
Sun (E_{max} = 2;13°)	70;52	2;5,39°		2;5°

From this we can conclude that Māshā'allāh's *Zīj al-Shāh* used the method of sines in computing the equations of the center; however, the values for the maximum equations for Saturn and Mars require slight adjustments. The new parameters will be:

Planet	*Argument*	*Method of sines*	*Text*
Saturn (E_{max} = 8;36°)	50;52°	6;40,2°	6;40°
Mars (E_{max} = 11;9°)	57;44°	9;25,40°	9;26°

7. For an explanation of these techniques and nomenclatures, see E. S. Kennedy, *Journal of Near Eastern Studies* 17 (1958), 112-121.

Bīrūnī

Planet	Argument	Method of sines	Method of declinations	Text
Saturn (E_{max} = 8;36°)	50°	6;35,17°	6;30,19°	6;35°
	51°	6;41,0°	6;36,11°	6;41°
Jupiter (E_{max} = 5;6°)	13°	1;8,50°	1;6,56°	1;9°
	29°	2;28,21°	2;25,0°	2;28°
Mars (E_{max} = 11;9°)	58°	9;27,21°		9;27°
Sun (E_{max} = 2;13°)	71°	2;5;45°	2;5,20°	2;6°
	80°	2;10,59°	2;10,51°	2;10°

Thus, Bīrūnī uses the same tables as does Māshā'allāh, but he finds the equations for degrees rounded off, whereas Māshā'allāh interpolates for minutes. Bīrūnī also gives one equation of the anomaly; 1;8° for Saturn for an anomaly of 171°. The maximum equation of the anomaly of Saturn according to the *Zīj al-Shāh* of Yazdijird is 5;44°; a table based on this parameter will have 0;59,36° as the entry opposite 171°. We do not, then, accurately control the equations of the anomaly for any of the planets, and all the longitudes computed in the tables given below are approximations only because of this.

It seems absurd, but is correct, for Bīrūnī to assume that Māshā'allāh computed the equations of the center from the true rather than the mean planetary longitudes; it should be noted that Bīrūnī himself makes the error of calculating the anomaly from the true, not the mean, longitude of the Sun. For the mean and true longitudes of Saturn and Jupiter on 18 March 333 according to Māshā'allāh's parameters, see below, pp. 85-86.

Having reconstructed the tables of the equations of the centers of Saturn, Jupiter, Mars, and the Sun in the *Zīj al-Shāh* used by Māshā'allāh, we must examine more closely the computation of their true longitudes. For the Sun, whose apogee is at ♊20° and whose maximum equation is 2;13°, it is easily determinable that the mean longitude for every year-transfer (λ = ♈0;1°) must be ♓27;49°; for then the equation of the center will be precisely +2;12°. This, of course, is what

Bīrūnī should have used in determining the anomaly of Saturn. The only other parameter for the Sun which is to be determined is its mean daily motion. Since the length of the year is 6,5;15,32,30^{d}, the mean daily motion is 0;59,8,10,0°.

Since we already have quite accurate values for the mean motions of Saturn and Jupiter and a table of mean longitudes for every horoscope (to be diminshed by 10° for all positions after the horoscope of the vernal equinox of −3360), as well as precise tables of their equations of the center, approximate tables of their equations of the anomaly, and exact values for the longitudes of their apogees, we should be in a position to reconstruct the method by which Māshā'allāh computed the whole series.

The following tables illustrate that we are indeed very near to a solution. Each equation is computed independently from the mean longitude of the planet; to use the Indian system of combining equations would, for the majority of cases, not affect the final result by more than half a degree.

Saturn

Date	*Text*	$\bar{\lambda}$	$\lambda = \bar{\lambda} + e(\bar{\lambda})$	*Text* − λ
−3380	♏ 1;56	♎ 23;43	♏ 1;33	+0;23
−3360	♊ 26;0	♊ 28;0	♊ 26;20	−0;20
−45	♐ 9;9	♐ 14;1−10	♐ 8;52	+0;17
−14	♑ 16;28	♑ 27;6−10	♑ 16;1	+0;27
333	♎ 9;8	♎ 10;34−10	♎ 8;17	+0;51
571	♏ 4;13	♏ 7;14−10	♏ 4;58	−0;45
609	♒ 9	♒ 21;18−10	♒ 6;2	+2;58
630	♏ 4;30	♏ 7;45−10	♏ 5;28	−0;58
650	♋ 2;2	♋ 12;2−10	♋ 0;52	+1;10
749	♏ 19	♏ 21;4−10	♏ 18;5	+0;55
829	♌ ⟨0⟩;6	♌ 8;7−10	♌ 0;14	−0;8
848	♓ ⟨1⟩5;15	♈ 0;12−10	♓ 12;46	+2;29
868	♐ 0;18	♐ 4;27−10	♐ 0;20	−0;2
888	♌ 2;16	♌ 8;42−10	♌ 0;53	+1;23
907	♓ 15;28	♈ 0;45−10	♓ 13;22	+2;6
928	♐ 12;1	♐ 17;15−10	♐ 11;42	+0;19

Jupiter

Date	*Text*	$\bar{\lambda}$	$\lambda = \bar{\lambda} + e(\bar{\lambda})$	*Text* − λ
−3380	♎12;39	♎12;33	♎13;1	−0;22
−3360	♊14;36	♊19;36	♊21;13	−6;37!
−45	♏28;58	♐6;56−10	♐2;8	−3;10!
−25	♋25;24	♌13;59−10	♋27;16	−1;52
−12	♍2;30	♍18;34−10	♍4;23	−1;53
333	♍22;44	♎18;19−10	♎8;20	−15;36!
571	♏4;26	♏12;2−10	♏5;11	−0;45
609	♑21	♑25;14−10	♑20;45	+0;15
630	♎23;4	♏2;31−10	♎24;31	−1;27
650	♊23;7	♋9;34−10	♊23;38	−0;31
749	♏6;8	♏14;0−10	♏7;21	−1;13
809	♏29;32	♐8;59−10	♐4;17	−4;45!
829	♋25;44	♌11;56−10	♋25;12	−0;32
848	♓11	♓18;40−10	♓11;33	−0;33
868	♏19;57	♏25;39−10	♏20;5	−0;8
888	♋18;20	♌2;37−10	♋15;51	+2;29
907	♓4;34	♓9;15−10	♓2;43	+1;51
928	♐12;11	♐16;38−10	♐12;14	−0;3

Thus we can compile the following statistics:

Difference	*Saturn*	*Jupiter*
Less than 1°	11	8
Between 1° and 2°	2	5
Between 2° and 3°	3	1
More than 3°	0	4
	16	18

Those cases in which the difference is greater than 1° may be instances where Māshā'allāh has deviated from our assumptions concerning his method of computing the equations or where either he or the scribe has made an error.

The foregoing is one possible solution to the difficulties of deriving parameters from these horoscopes. It involves the assumption of an error of 10° in the mean longitudes of Saturn and Jupiter in the computation of the series of their mean conjunctions and of the horoscopes for –3380 and –3360. In any case, these two dates – –3380 and –3360 – are firmly fixed as those for which Māshā'allāh intended to cast the horoscopes; this is proved by the intervals given both by him and by others between the various events in his chronology and by the fact that the cycle of Mighty Fardārs began in –3380.

Another solution, however, has recently been offered by J. J. Burckhardt and B. L. van der Waerden.[8] They, being unaware of the evidence for the dating of horoscopes 1 and 2, date them in –3320 and –3300 respectively. For Saturn and Jupiter 60 years involve a mean motion of about 13° and 20° respectively. If the corrective factor of 10° is applied, there are differences of only 3° and 10° which are to be distributed over more than 3000 years since the next horoscope in the series is securely dated –45. The differences between the mean motions of Saturn and Jupiter under the two solutions, then, will be minute indeed. It is doubtful whether Māshā'allāh's computations or Ibn Hibintā's and his scribe's numbers are correct enough to permit a judgment between the two solutions.

It might be argued that the lunar node and Mars provide a satisfactory check. But the calculation of the mean yearly motion of the node given above shows that the difference of 60 years at the beginning of the series of horoscopes does not preclude the derivation of a reasonable parameter from the material at our disposal.

The mean motion of Mars is more difficult to control, since we have given no mean longitudes for that planet. The situation is made particularly difficult by the insecurity of the textual entries; thus there are serious variants in the horoscopes of –3380 and 609 and substantial errors in that of 333 and in horoscopes 9 and 13 of the *Fī qiyām al-khulafā'*. With its enormous epicycle, Mars is always subject to more serious computational errors than are Saturn and Jupiter, especially when the computer is as incompetent as was Māshā'allāh.

And again the long period of more than three millennia between the second horoscope and the third produces a situation wherein very slight modifications of the mean yearly motion will radically change the true longitudes of the planet.

8. "Das astronomischen System der Persischen Tafeln I," *Centaurus* 13 (1968), 1-28.

If, for instance, as the converse of this, we take an interval between two horoscopes for which the true longitudes (and therefore also the mean longitudes) of Mars are close together, we will find little difference between the mean yearly motions. In horoscope 1 the longitude of Mars is ♎27;17, in the horoscope of 848 it is ♏8;12. If horoscope 1 is dated –3380, the mean yearly motion is 3,11;24,44°; if it is dated –3320, the mean yearly motion is 3,11;24,12°. Considering the crudeness of the method, a difference of 0;0,36° is not very significant. Applying the method to the positions of Mars in the horoscopes of –45 and 888 one finds a yearly mean motion of 3,11;23,27°; or between –45 and 809, 3,11;23,48°; or between –3360 and 868, 3,11;24,31° (if horoscope 2 is dated –3300, the mean yearly motion is 3,11;24,1°).

It is true that, for the interval between –45 and 928, a difference of about 0;0,20° in the mean yearly motion of the planets would shift the mean longitudes by about 5°. The mean yearly motions resulting from a dating of horoscopes 1 and 2 in –3380 and –3360 seem to be at least 0;0,20° higher than the mean yearly motion adopted by Burckhardt and van der Waerden (3,11;24,4°). However, it is unclear how the equations, which we do not control, affect this result, and the text upon which the result rests is not secure. Still, this would be a telling argument in favor of the hypothesis of Burckhardt and van der Waerden were it not for the fact that their solution for Mars is the least satisfactory of all their solutions. Their results are within less than 1° of the text's entries in only 7 out of 15 cases, and are more than 3° from the text's entries in 6 cases.

The choice, then, lies between assuming that Māshā'allāh computed horoscopes 1 and 2 for –3380 and –3360, but made an error of 10° in calculating their mean longitudes, or that he intended to compute the horoscopes for those dates, but by mistake computed them for –3320 and –3300. In favor of the latter hypothesis is the situation with regard to Mars, but that is questionable enough not to be decisive.

4. COMMENTARY

f214v

8-10. See also by Māshā'allāh: (1) *Liber Messahallae de revolutione annorum mundi*, ed. J. Heller (Noribergae, 1549), (2) *Epistola Messahalae de rebus eclipsium et de coniunctionibus planetarum in revolutionibus annorum mundi*, translated into Latin by Ioannes Hispalensis and edited by J. Heller (Noribergae, 1549), and by N. Pruckner in *Iulii Firmici Materni Astronomicῶn Libri VIII* (Basileae, 1551), pp. 115-118, as *Messahallach de ratione circuli et stellarum et qualiter operantur in hoc seculo.* For the Hebrew translation by Abraham ben Ezra, see now B. Goldstein, "The Book on Eclipses of Māshā'allāh," *Physis 6* (1964), 205-213.

The great Arabic bibliographer Ibn al-Nadīm mentions among Māshā'allāh's works a *Kitāb fī al-qirānāt wa al-adyān wa al-milal (Fihrist*, Cairo, n.d., p. 396). It is presumably from this work that Ibn Hibintā has drawn.

13. There are, of course, many works on astrology in Greek, Latin, and Arabic ascribed to Hermes; the only one I know of in which the conjunctions and transfers of world-years are discussed is in Par. ar. 2487 f.36. However, Sahl ibn Bishr (Vat. ar. 955 ff. 173v-175v) quotes a year-transfer horoscope from a book of Hermes. "Other astrologers" who have discussed the subject in Persian and Arabic are extremely numerous, and need not be catalogued here.[1]

17. This (mean) conjunction of Saturn and Jupiter can be dated 3 November A.D. −5782; for this and other dates see the note on chronology.

1. For a partial enumeration see E. S. Kennedy, *Actes du dixième congrès international d'histoire des sciences* (Hermann, Paris, 1964), 1:32-37.

18-19. The midpoint of the millennium of Mars was 2 August −5791.

19-215r:1. The choice of a conjunction in Taurus was probably influenced by the role played by the sacrifice of the Primordial Bull in the creation of the material world.

f215r

1. There were nine more mean conjunctions in the triplicity of the earth.

2-4. Parameters closer to those actually used by Māshā'allāh, would be ⟨24⟩2;25,35° and 19^{y} 10^{m} 10^{d} 10^{h} (=2,0,50;20,17,30^{d}).

5. The 121st mean conjunction marked the transfer from the triplicity of air to that of water (which naturally signifies the Deluge); it occured on 25 September −3380. Māshā'allāh's figure of 2412^{y} 6^{m} 26^{d} has been discussed in the note on chronology.

9-10. The horoscopes investigated and interpreted by Māshā'allāh are those of the vernal equinoxes of the years in which mean (later true) conjunctions of Saturn and Jupiter took place. The return of the Sun to Aries is called a revolution of the world-year. See, for example, *De revolutione annorum mundi*, preface: "Et patefaciat tibi [Deus] revolutionem annorum, cuius scientia et dispositio est ut scias quando ingrediatur Sol in primum minutum signi Arietis."

f215v

Horoscope 1 is dated 11 February −3380

	Text		Text
♄	♏ 1;56 retr.	☿	♓ 3;24
♃	♎ 12;39 retr.	☾	♉ 23;13
♂	♎ 27;17	H	♐ 19
☉	♈ 0;1	☊	♋ 8
♀	♓ 26;16		

A copy of this horoscope is found in the *Kitāb al-qirānāt wa taḥāwīl sinī al-'ālam* of al-Sijzī.[2]

2. See D. Pingree, *The Thousands of Abū Ma'shar* (Studies of the Warburg Institute, 30; London, 1968), p. 81.

♄	♏ 1;55 retr.	☿	♓ 25
♃	♎ 12 retr.	☾	♉ 13;13
♂	♎ 6;45 retr.	H	♌ 2;25
☉	♈ 0;0	☊	♋ 8
♀	♓ 27;16		

The date for the year-transfer of the conjunction indicating the Flood – the vernal equinox of –3380 – was previously computed from the Abū Ma'shar tradition by a scholar in the mid-fourteenth century; see Appendix 5.

1-3. (a) Jupiter is lord of Sagittarius, the ascendant;

(b) Sagittarius is of the triplicity of fire, whose diurnal lord (but the Sun is below the horizon) is the Sun;

(c) The Sun "confers its power" by aspecting Jupiter diametrically;

(d) Venus also aspects Jupiter diametrically;

(e) Venus is lord of Taurus, the sign in which the Moon is;

(f) Venus in ♓26;56 can be said to be in her exaltation in ♓27.

Therefore Jupiter, being itself strong and being aspected by two strong planets – the Sun and Venus – is lord of the year.

Cf. *De revolutione annorum mundi.* cap. 5: "De inveniendo dominum anni. Cum ergo volueris scire dominum anni, aspice hora revolutionis planetam qui fuerit fortior ceteris in loco suo et plura habuerit testimonia, ac pone eundem dominum anni *in climate in quo fueris.*"

Chapter 6 of the same work gives rules for determining which planet is strongest:

(a) "Scito quod fortior ex planetis est ille qui fuerit in ascendente non remotus ab angulo neque cadens vel qui fuerit sic in medio coeli.

(b) "In occasu vero et angulo terrae erit inferius quam dixi tibi in fortitudine per quartam partem. . .

(c) "Cum fuerit dominus ascendentis in horoscopo, scilicet per tres gradus ante vel retro cuspidem eius, non cadens neque remotus ab angulo ascendentis, non erit nobis necesse cum eo aspicere alium planetam.

(d) "Similiter dominus exaltationis signi ascendentis cum fuerit in gradibus exaltationis suae.

(e) "Dominus vero triplicitatis cum fuerit in ascendente habebit tertiam partem fortitudinis domini ascendentis.

(f) "Dominus quoque termini habet quintam partem fortitudinis, et hoc secundum fortitudinis eorum in signis.

(g) "Dominus horae habet septimam partem.

"Et scito quod hoc fit cum fuerint in ascendente vel medio coeli. Si autem fuerint in occidente vel in angulo terrae minuetur eorum fortitudo; similiter cum fuerint in undecimo, nono, et quinto magis ea minuetur."

Very relevant to the present horoscope is chapter 10; "Cum commiserint luminaria dispositionem suam domino ascendentis aut alterum eorum, non erit tibi necesse aspicere cum hoc alium quia ipse est dominus anni absque dubio."

3-5. There are two methods of reckoning the twelve places or δωδεκάτοπος. The simplest ("by counting") merely counts each sign as a place; the other ("by division") uses equal arcs of 30° along the ecliptic measured from the ascendant degree, which naturally appear as unequal to observers on the surface of the earth. Since these horoscopes indicate world events, it might be reasonable to suppose that Māshā'allāh employed tables of rising-times in sphaera recta – that is, he places the observer at the equator rather than at any other arbitrary latitude. But I assume that he used a simple Babylonian System A of rising-times, as good for Baghdad as for Babylon; for the justification see 221v:1-3. It is clear that the Moon is in the sixth place by counting, Saturn in the twelfth, and Jupiter and Mars in the eleventh. But by division Jupiter is 239° from the ascendant, or just a degree before the beginning of the tenth place; and Mars is 20° within the tenth.

5-6. On the evil effects of retrogressions see *De revolutione annorum mundi*, chapters 20 and 25.

6. The tenth sign, in which all three superior planets are by division, is the "upper midheaven"; Libra, in which Jupiter and Mars are, is a human sign. The other human signs are Gemini, Virgo, the first half of Sagittarius, and Aquarius; see 227r:12-13.

f226r

1. The moving signs are Aries, Cancer, Libra, and Capricorn; the fixed signs their *ἐπαναφοραί* Taurus, Leo, Scorpio, and Aquarius; and the common signs or signs of two bodies the moving signs' *ἀποκλίματα* Pisces, Gemini, Virgo, and Sagittarius. As the horoscope is nocturnal, the Moon is the

luminary which is its lord. In fact, by signs the Moon is aspected diametrically by Saturn, which should injure it; however, Māshā'allāh may insist here on an aspect within 15°.

8. Mean conjunction 122 occurred on 31 July −3360.

10-12. The horoscope is dated 12 February −3360; the Deluge therefore occurred in June–July of that year. The Sun was in its own house, Leo, at the time of the Deluge.

f226v

Horoscope 2

Text

♄	♊ 26;0	☿	♓ 3;22
♃	♊ 14;36	☾	♍ 29;30
♂	♉ 19;49	H	♓ 4
☉	♈ 0;1	☊	♊ 11
♀	♈ 1;2		

This horoscope also is found in al-Sijzī's work mentioned in the commentary to 215;9-10.[3]

♄	♊ 27	☿	–
♃	♊ 14;36	☾	♈ 26;35
♂	♉ 17;47	H	♓ 5
☉	♈ 0;0	☊	♊ 11
♀	♈ 1;3 retr.		

1ff. Māshā'allāh compares Horoscopes 1 and 2.

3-5. Jupiter is lord of this year, also, because:

(a) It is lord of the ascendant, Pisces;

(b) According to Ptolemy, Jupiter's term in Gemini is 8°-13°, in Pisces 9°-14°; however, the fact that the horoscope is nocturnal must be understood to make his term in Pisces 0°-8°, which therefore includes the ascendant degree;

(c) Jupiter is nocturnal lord of the triplicity of fire, which includes Aries;

(d) The exaltation in Pisces (at 27°) belongs to Venus;

(e) The Sun and Venus aspect Jupiter in sextile.

Jupiter is in Gemini, the fourth place or hypogee ("cardine of the earth") by counting.

3. Pingree, *The Thousands of Abū Ma'shar*, p. 82.

f227r

1. Jupiter is not lord of Gemini or of its triplicity nor is its exaltation in Gemini.

1-2. Mercury, the lord of Gemini, is in Pisces, the ascendant.

3. The heliacal rising of Saturn should occur in the fourth month, when the Sun is in Cancer. The evil planet Saturn attains great strength at its heliacal rising, and thereby precipitates the cataclysm; see *De revolutione annorum mundi.* chapter 38: "Scito quod planeta cum egreditur de subradiis et combustione erit sicut puer, qui proficit et augmentatur."

3-6. Jupiter will be in Cancer at the time of the mean conjunction, of course, and this is the fifth sign from the ascendant. Ordinarily this would be interpreted as indicating five years; but, as the Moon is lord of Cancer and is in a cardine (Virgo, the descendant), it signifies months. One might argue that the Moon in a cardine should be lord of the year rather than Jupiter, especially as the horoscope is nocturnal; see *De revolutione annorum mundi*, chapter 7.

6-15. See *Epistola*, chapter 8.

14. The animal signs are Aries, Taurus, Cancer, Leo, Scorpio, second half of Sagittarius, Capricorn, and Pisces.

16. It is unclear why the longitude of the conjunction should be termed the "lot of the transit."

17-19. See *Epistola*, chapters 8 and 9.

f227v

1-4. The shifts indicate what happens while the conjunctions remain within one triplicity (that is, for periods of 238/9 or 258/9 years); the conjunctions, what happens between themselves (that is, for 19/20 years); and the transfer of the world-year, what happens during each solar year. See, for example, Abū Ma'shar, *Kitāb al-qirānāt*.

9. 2423^y 6^m 12^d is the time Ibn Hibintā arrives at by computation as having elapsed between mean conjunction 0 and mean conjunction 122.

11. The triplicity of fire was entered on mean conjunction 282 with a mean conjunction at the beginning of Aries.

12. The eighth mean conjunction of this triplicity – the 289th of the series – occurred on 1 August –45.

14-15. The 5749^y 5^m 8^d, as I have shown in the note on chronology, is the time computed by Māshā'allāh as having elapsed between the midpoint of the millennium of Mars and mean conjunction 289.

16. This horoscope is dated 13 March –45.

f 228r

Horoscope 3

	Text	*Computation*	*Text – computation*
♄	♐ 9;9	♏ 28	+11
♃	♏ 28;58	♏ 24	+5
♂	♒ 10;49	♒ 0	+11
☉	♈ 0;1	♓ 20	+10
♀	♒ 15;29	♒ 7	+8
☿	♈ 9	♒ 23	+36
☾	♈ [2]6;13	♈ 3	+3
H	♈ 10;37		
☊	–		

1-2. The Sun being in the ascendant, Aries, in which is its exaltation (at Aries 19°), is declared to be lord of the year in accordance with *De revolutione annorum mundi*, chapter 7. The only disturbing factor – not mentioned by Māshā'allāh – is that this is a nocturnal birth, and it therefore might be better to make the Moon, which is also in Aries, the lord of the year.

2-4. Saturn, in Sagittarius, in the ninth place (the place of religion in the δωδεκάτοπος), is aspected in trine by the Sun, the Moon, and Mercury from Aries; this combination of age, brilliance, wisdom, and so forth in the place of religion naturally leads to the birth of a prophet. In theory, what really should cause the birth of a prophet is a conjunction in the tenth place; see *Epistola*, chapter 10: "Item scias quod coniunctio maior cum fuerit in aliquo angulorum et maxime in angulo medii coeli, significabit apparitionem regis vel prophetae ex parte eiusdem signi." The conjunction will be in the tenth place from the year-transfer's ascendant by division.

4-5. The slowness of Saturn presumably accounts for the delay, and the nature of Sagittarius (it is two-bodied) explains why the event is delayed to the next conjunction.

6. Trine aspect is always favorable.

6-228v:1. The lord of Sagittarius, the sign occupied by Saturn, is Jupiter; and Jupiter is in the twelfth place from it. It is unclear why this should indicate the crucifixion.

f 228v:1. The ninth mean conjunction in this triplicity is number 290. It is dated 9 June −25.

3. This horoscope is dated 14 March −25.

Horoscope 4

	Text	*Computation*	*Text − computation*
♄	⟨♌⟩	♋ 29	
♃	♋ 25:24	♋ 21	+4
♂	♋ 20	♋ 13	+7
☉	⟨♈0;1⟩	♓ 21	+9
♀	–	♈ 21	
☿	⟨♈⟩	♒ 26	
☾	♍ 3	♍ 2	+1
H	♋ 1		
☊	–		

5-229r:4. The Sun, "the luminary of the day" (the horoscope is diurnal), and Mercury, the lord of Virgo, the sign occupied by the Moon, both aspect Saturn in trine; it will be recalled that in Horoscope 3 it was the trine aspect of Saturn in the ninth place by the Sun, the Moon, and Mercury that determined that a prophet would be born. The seventh and eighth places from the ascendants are Capricorn and Aquarius, the two houses of Saturn; Saturn, in Leo, is thirteen signs from the ascendant.

f229r:5. This horoscope is dated 14 March −12.

Horoscope 5

	Text	*Computation*	*Text − computation*
♄	♑ 16;28	♑ 7	+11
♃	♍ 2;30 retr.	♌ 29	+3
♂	♊ 21;22	♊ 18	+3
☉	♈ 0;1	♓ 22	+8
♀	♉ 14	♉ 7	+7

☿	♈ 4;59	♒ 26	+39
☽	♊ 22;44	♊ 26	
H	♎ 18		
☊	–		

This horoscope is copied in al-Sijzī's work,[4] in the *Ta'rīkh al-Ya'qūbī*,[5] and in the *Dastūr al-Munajjimīn*[6]:

	al-Sijzī	*al-Ya'qūbī*	*Dastūr*
♄	♑ 16;28	♑ 16;28	–
♃	♍ 2;31 retr.	♍ 1;30 retr.	2;31
♂	♊ 21;22	♊ 21;44	21;32
☉	♈ 0;0	♈ 0;1	–
♀	♉ 14	♉ 14	14
☿	–	♈ 4;17	17(?)
☽	♊ 22;4	–	21;25
H	♎ 18	♎ 18	7;18 or 4;38
☊	♈ 25;15	–	–

f 229v: 1-4. Saturn is the lord of the year because:

(a) All the planets except Mars and the Moon aspect it; (1) Jupiter in trine, (2) the Sun and Mercury in quartile, (3) Venus in trine;

(b) It is in its own house, Capricorn;

(c) Its exaltation, Libra (21°), is in the ascendant, and it is diurnal lord (though it is a nocturnal horoscope) of the triplicity of air to which Libra belongs;

(d) The lord of Libra, the ascendant, is Venus and it is in its other house, Taurus, aspecting Saturn in trine.

4. The ten months represent the ten signs between the Sun and Saturn; they place the birth of Christ in December.

5. The violence Christ will meet – the crucifixion – is here predicted from the fact that, in a nocturnal horoscope, a nocturnal planet, the Moon, is in conjunction with the maleficent planet Mars in the ninth place, the place of religion. I do not understand this.

4. Pingree, *The Thousands of Abū Ma'shar*, p. 81.
5. Edited in 2 vols. (Beirut, 1960), vol. 1, p. 68.
6. M. S. Arabe 5968 in the Bibliothèque Nationale, Paris, f. 274v.

5-6. "Fear. . . of death in a high place" must be the temptation to jump from a high place (Matthew 4:5-7; Luke 4:9-12). It is predicted because the Moon and Mars are in conjunction in "the highest of the signs"; but this is not strictly accurate, as the *μεσουράνημα* is in Cancer by both counting and division.

6-19. The errors in Ibn Hibintā's computations have been discussed in the note on chronology.

f230r:1. Mean conjunction 320 occurred on 12 December 570.

2. The Prophet Muḥammad was born in c. 571; for Māshā'allāh's dating of the event, see below, p. 127.

3-6. Bahman is the eleventh month in the Persian calendar. If one simply extended the calendar of Yazdijird backward, the Persian year in which the vernal equinox of A.D. 571 occurred would have begun on 2 July 570; then 22 Bahman would have been 19 May 571. As Ibn Hibintā remarks, there are no years given to use as a base; but we can date the horoscope of the vernal equinox to 19 March 571. If 22 Bahman = 19 March 571, then 1 Farvardīn was 3 May 570 – exactly 60 days before 2 July. This indicates that two Persian months of 30 days each were intercalated between the inauguration of the calendar here used by Māshā'allāh – it is that of the *Zīj al-Shāh* written for Kisrā Anūshirwān in 556 – and the calendar of the *Zīj al-Shāh* written for Yazdijird III, whose epoch was 16 June 632.

Horoscope 6

	Text	*Computation*
♄	♏6;13[7] retr.	♏6 retr.
♃	♏4;26 retr.	♏6 retr.
♂	♊12;15	♊16
☉	♈0;1	♈1
♀	♈2;16	♈3
☿	♈18;26	♈1 retr.
☾	♋1;10	♋1
H	♎22	
☊	⟨♏ 7;53⟩	

7. The position of Saturn must be changed to ♏4;13; see the note on chronology, pp. 79-80.

We have also, besides numerous versions of this horoscope by other astrologers, two further copies of Māshā'allāh's diagram, one in al-Ya'qūbī (vol. 2, p. 7), the other in *Dastūr* (Paris BN Ar. 5968, f. 309):

	al-Ya'qūbī	*Dastūr*
♄	♏ 6;23 retr.	♏ 6;53
♃	♏ 3;23	♏ 3;23 or 4;26
♂	♊ 12;15	♊ 12;15
☉	♈ 0;0,1	♈
♀	♈ 1;56	♈
☿	♈ 18;16	♈
☾	♋ 1;20	♋ 1;5
H	♎ 22	♎ 25;45
☊		

f 230v: 1-6. Normally, the "Mighty Intihā' " travels one sign every millennium. Now, if we began with Aries in the millennium of Mars (in the middle of which occurred mean conjunction 0), the Deluge happened at the end of the Intihā' of Gemini and Muḥammad was born in the Intihā' of Libra. A cycle of "Mighty Fardārs" begins with Saturn and Cancer in –3380.[8] Each subsequent Fardār, consisting of 360 years, is associated with the next planet and sign in order. Therefore, the Fardār of Gemini (of which Mercury is lord) and Venus commenced in A. D. 580. Our astrologer connects this with the mean conjunction of 571, and states correctly that this Fardār of Venus and Mercury will last 360 years.

If we assume that the division (*qisma*)–which is also characteristically used in genethlialogy[9] – began its motion at Aries at conjunction 0, then in the 2423 years Māshā'allāh assigns between conjunctions 0 and 122 it travelled to Sagittarius 23°. Now Māshā'allāh further puts 3912 years between conjunctions 122 and 320; but this places the division in Scorpio 15° rather than in Pisces 20°.

8. Pingree, *The Thousands of Abū Ma'shar*, p. 60 n. 5.

9. See the anonymous Byzantine translations of Abū Ma'shar's *Kitāb aḥkām taḥāwīl sinī al-mawālīd*, ed. David Pingree (Leipzig, 1968), book 3.

However, realizing that the periods of a Mighty Fardār and of a complete revolution of the division are equal, we can say that, if the Mighty Fardār of Gemini and Venus began in 580 with the division in Pisces 20°, 11 periods previously – in –3380 – it was again in Pisces 20°. This, of course, puts it in Aries in the year of the Deluge according to Māshā'allāh, –3360; however, in –3101, the year of the Deluge according to Abū Ma'shar and the beginning of the present Kaliyuga, the division would have been in Sagittarius 25°. Therefore, this reference would appear to have been inserted by Ibn Hibintā.

6-16. The intervals in days used by Ibn Hibintā are, respectively,

Vernal equinox of 571 to Hijra	18713^d 16^h
Hijra to Yazdijird	3624^d
Vernal equinox of 571 to Yazdijird	22342^d 16^h.

The discrepancy of 5^d represents the epagomenal days which are dropped in adding 11^m (330^d) and 3^m (90^d) to get 1^y (365^d) 2^m (60^d). Since Horoscope 6 is an evening horoscope, the 18713^d 16^h will bring us to about noon of 13 June 622, which is, of course, the wrong date for the beginning of the Hijra. The interval of 3624^d between Hijra and Yazdigird is correct, and can be found in many sources.

Ibn Hibintā has said before that from conjunction 0 to conjunction 122 (Deluge) is 2423^y 6^m 12^d and that from conjunction 0 to conjunction 320 is 6336^y 0^m 26^d; simple subtraction gives the interval from conjunction 122 to conjunction 320 as 3912^y 6^m 14^d. However, if one adds to this the interval given between the vernal equinox of 571 and the beginning of the Era of Yazdijird – 61^y 2^m 17^d – one gets 3973^y 9^m 1^d instead of Ibn Hibintā's 3974^y 9^m 1^d. Of course, all the computation is meaningless. Abū Ma'shar's interval of 3733^y 2^m $26;9,10^d$ between his date of the Deluge (17/18 February –3101) and the beginning of the Era of Yazdijird appears not to be that given in the *Kitāb al-ulūf* as stated by al-Hāshimī (1,363,598^d = 3735 Persian years and 323^d) or in *Kitāb al-qirānāt* 4,12 (3735 Persian years and 322^d)[10]. In fact, the *Kitāb al-ulūf* figure of

10. *The Thousands of Abū Ma'shar*, p. 40.

$1,363,598^d$ is correct; if this equals 3733^y 2^m $26;9,10^d$, then the year-length is $6,5;15,32,24^d$, a parameter attested elsewhere for Abū Ma'shar's "Persian System."[11]

17-231:2. Saturn is lord of the year because:

(a) It is in the ascendant by division;

(b) It is lord of (that is, near) its exaltation in Libra 21°;

(c) The Moon, which is lord of the night (and this is a nocturnal horoscope), aspects it in trine from the tenth place by counting, Cancer. By division, the Moon is in the 14th degree of the eighth, *not* of the ninth, place; Māshā'allāh probably computed from Libra 0° rather than from the true ascendant, Libra 22°. The Moon is aspected by Jupiter in trine. It should be noted as well that the Moon is in its own house.

f231r:2-4. Mars in Gemini is in the ninth place by counting, close to being in the *μεσουράνημα*, "in the highest part of the orbit." It receives sextile aspect from the three planets in Aries – the Sun, Venus, and Mercury.

4-6. The "first lot" is somewhat mysterious. However, the lord of Scorpio, Mars, is in the ninth place, Gemini, which indicates religion; and therefore a religious prophet is predicted. See also *Epistola*, chapter 10, quoted on 228r:2-4. This conjunction is in the ascendant by division. The two years result from Saturn's being in the second place by counting. Venus' involvement through her quartile aspect of the Moon, the significator regis as lord of the tenth place, indicates that the prophet will be born at Tihāma in the land of the Arabs; for Venus is the star of that nation.

7-8. Cf. 228r:4-5. If Saturn were in Sagittarius, it would be aspected in trine by Venus and diametrically by Mars.

8-11. The Moon's being in its own house, Cancer, in a cardine, that of midheaven, is a good sign; it is, of course, the significator regis and benefits the Prophet. However, the hardships of the early part of Muḥammad's ministry are to be predicted from the fact that Venus, the star of the Arabs, has just left conjunction with the Sun and is still under its rays.

11-12. Mars in the ninth place, that of religion, indicates the violence which will accompany the establishment of Islām.

11. *Ibid.*, pp. 30 and 36.

13-15. The Battle of the Elephant, in which the Ethiopian army was defeated by the combined Persian and Arab forces under Vahrīz, took place in 570.

16-19. Indeed (19^y 10^m 10^d) X 3 = 59^y 7^m; but, as always, this is a meaningless statement. The 1^y 9^m 17^d 16^h might be a mistake for 1^y 7^m 17^d 16^h; for this plus 59^y 7^m equals 61^y 2^m 17^d 16^h, the time given in 230r:10-11 as the interval between the vernal equinox of 571 and the beginning of the Era of Yazdijird. But then, since the Era of Yazdijird began on 16 June 632, this computation puts the mean conjunction on 1 November 630; and 1^y 9^m 18^d before 16 June 632 is 2 September 630. But, in fact, mean conjunction 323 fell on 4 July 630. The 60-day difference between 2 September and 4 July 630 is undoubtedly due to Māshā'allāh's continued use of Anūshirwān's calendar, in which 1 Farvardīn of the year in which Yazdijird III began to rule fell on 16 April 632.

f231v:2-3. This horoscope is dated 19 March 630.

Horoscope 7

	Text	*Computation*
♄	♏ 4;30	♏ 8 retr.
♃	♎ 23;4	♎ 26 retr.
♂	♑ 5	♑ 5
☉	♈ 0;1	♈ 1
♀	♓ 19	♓ 19
☿	♈ 11;13	♈ 8
☾	♈ 2;30	♈ 1
H	♑ 18	
☊	♓ 1	

A copy of this horoscope is also preserved by al-Ya'qūbī (vol. 2, p. 113):

al-Ya'qūbī

♄	—	♀	17
♃	♎ 23;4 retr.	☿	11;13
♂	♑ 0;5	☾	♈ 2;30
☉	♈ 0;1	H	♑ 18
		☊	

Moreover, we have from al-Ya'qūbī (vol. 2, p. 22) and the *Dastūr* Māshā'allāh's horoscope for the year-transfer of 609 (his date for the third conjunction of this shift). The longitude of the Moon forces us to date it 26 March 609; it should, of course, be for 19 March. This date would put the Moon in Cancer 17°.

	al-Ya'qūbī	*Dastūr*	Computation (19 March 609)
♄	♒ 9	♒	♒ 12
♃	♑ 21	♑	♑ 23
♂	♍ 13 retr.	♍ (?)	♍ 14 retr.
☉	♈ 0;1	♈ 0	♈ 2
♀	♓	—	♓ 0
☿	♈ 14	♈ 14(?)	♈ 19
☾	♎ 17	♎ 17	♋ 17
H	♍ 4	♍	
☊			

4. The "base sign" is here, of course, Scorpio.

5-6. One maleficent planet, Mars, is in the ascendant cardine by counting; the other, Saturn, is in the tenth place by division.

6-232r:1. Saturn is strong because it is lord of the ascendant (Capricorn), is near its exaltation (in Libra 21°), and is aspected (a) by Mars in sextile, (b) by Venus in trine.

f232r:1-2. Scorpio indicates the Arabs, and Saturn their victories; see 232r:6.

2. Mars aspects Saturn in sextile and its exaltation is in 28° of Capricorn, the ascendant sign.

3. Mars receives Saturn because Saturn is in Mars' house; see *Liber Messahalae de receptione*, ed. J. Heller (Noribergae, 1549), chapter 1.

3-5. Māshā'allāh elsewhere (*De revolutione annorum mundi*, chapter 3) assigns the fourth climate (Babylon) to the Sun; Saturn gets the first and Mars the third. The choice of the fourth climate in the present horoscope is dictated by the following rule (*ibid.*, chapter 11): "Regi vero climatis iudicabis ex domino medii coeli (here Mars, lord of Scorpio) et a Sole et ex coniunctione ac separatione planetarum ab eis; et eliges ex Sole et domino medii coeli fortiorem secundum fortitudinem locorum ponesque vincentem significatorem regis."

Between Mars and the Sun there is little choice as to comparative strength. Each is in the sign of its exaltation. However, Mars is also in the ascendant and is aspected in sextile by Venus and Saturn. Therefore the king goes East (the ascendant's direction) before meeting his death; but the climate is determined by the Sun. Mars, as significator regis, aspecting Scorpio, in which Saturn is, should determine that the "victory of the evil people" came from the third climate; *ibid.*, chapter 14: "Aspice ipsam domum, in qua fuerit malus planeta, utrum significator regis vel dominus anni habuerit in ea aliquod testimonium; quod si evenerit, erit inimicitia ex eius climate secundum testimonii illius quantitatem" (the last phrase makes it very slight). However, chapter 29 gives further rules for determining the enemies of the king: "Et si (malus) aspexerit (significatorem regis) ex domo significatoris regis, erit hostis ex domesticis eius; et si eum aspexerit ab exaltatione sua, erit ex nobilibus, de his qui fuerint ei consimiles et eiusdem ordinis . . . et scito quod trinus et sextilis aspectus significat propinquos." For the "disputes and shedding of much blood," *ibid.*, chapter 45: "Et si aspexerit Mars significatorem regis et fuerit commixtus lumini Saturni, significat contentionem et effusionem sanguinis." For the "general misery and violence," *ibid.*, chapter 14: "Malus quoque cum fuerit in medio coeli et impedierit dominum anni, intrabit hoc (malum) inter omnes homines."

Scorpio, as we said before, is the sign of the Arabs, and Saturn is the lord of the year; we read *ibid.*, in chapter 16: "Aspice signum in quo fuerit significator anni cuius climatis sit vel cuius civitatis, quia eiusdem climatis dominus erit fortior et successus habebit foeliciores prae ceteris climatibus." Venus, moreover, the star of the Arabs, is near its exaltation (it is in Pisces 19°, its exaltation Pisces 27°) and is in conjunction with the ascending node of the Moon (in Pisces 1°), a favorable indication; *ibid.*, chapter 24: "Scito quod Caput Draconis congruit fortunis."

These "predictions" do in fact describe tolerably well the troubles of the Sasanian Empire between the murder of Kisrā II in 628 and the accession of Yazdijird III in 632, briefly chronicled in Christensen, *L'Iran sous les Sassanides*, 2nd ed. (Copenhague, 1944), pp. 497-499. In this short period of four years there reigned ten monarchs:

Kavādh II Shērōē (six months; poisoned ? died of plague?);

Ardashīr III (one year and six months; murdered by the general Farrukhān Shahrvarāz);

Shahrvarāz (murdered by nobles);

Kisrā III (recognized as Shāh in Eastern Iran; assassinated by the governor of Khurāsān);

Bōrān (one year and four months);

Pērōz II;

Āzarmēdukht (murdered by Rōstahm);

Hormizd V;

Kisrā IV;

Farrukhzādh-Khusrō.

But it seems that only one Shāh, Kisrā III, was slain in the East, and he is surely too insignificant to be referred to by Māshā'allāh. Rather, one suspects that the tragic end of Kisrā II has been postdated a few years (and combined with that of Yazdijird III ?). Heraclius in 628 had taken the royal residence of Dastgard and threatened Ctesiphon; Kisrā II did not flee to the East as indicated by Mars in the ascendant, but he was slain through a conspiracy of generals, nobles, and his own son, in Ctesiphon.

The fantastic success of Arab arms in "seeking the kingdom" between 630 and 650 need not be detailed here.

7-9. Saturn aspects the ascendant, Capricorn, its own house, in quartile, a bad aspect; therefore there is "fear of the people of the house of the Prophet" in the third year. In fact, Muḥammad died in the third year – on 8 June 632; but I fail to see the astrological necessity of this.

9-15. The horoscope is indeed nocturnal, and the Moon, only 2;30° from the Sun, is under its rays; furthermore, the Sun is unfortunate because of the quartile aspect of Mars, and both the luminaries are cadent by division. The bad situation of the Sun indicates evil for the fourth climate, Babylon.

Al-Madā'in refers to the two cities of Seleucia (Vēh-Ardashīr) and Ctesiphon, respectively on the right and left banks of the Tigris. The Sasanian capital was taken by Sa'd ibn Abī Waqqāṣ in June 637. This was indeed after four years and a fraction of Yazdijird's rule; it was also three years after Khālid ibn al-Walīd's taking of al-Ḥīrah and the first Arab incursions into 'Irāq.

15. Mean conjunction 324 occurred on 11 May 650.

17. Horoscope 8 is dated 19 March 650.

	Text	Computation
♄	♋2;2	♋2
♃	♊23;7	♊26
♂	♊22;40	♊17
☉	♈ 0;1	♈ 2
♀	♉11:23	♉10
☿	♈ 4:40 retr.	♈ 0 retr.
☾	♌16:17	♌11
H	♈10	
☊	⟨♒⟩	

f233v:2. The Sun is lord of the year because it is (a) a luminary in the ascendant (by counting), (b) in the sign of its exaltation (Aries 19°), (c) aspected (1) by Jupiter and Mars in sextile, (2) by Saturn in quartile, (3) by the Moon in trine.

3-13. The Sun again indicates the fourth climate, Babylon, and its aspecting of Saturn in quartile foretells the death of the King of Babylon (the Sasanian Emperor) in the fourth year (one year for each sign between the Sun and Saturn). However, Yazdijird III was assassinated near Marw in 651 or 652, not in 653.

In nine years Saturn will travel c. 108° to Libra 20°; its exaltation is in Libra 21°; and it will then be in opposition to the Sun in Aries. This configuration is taken by Māshā'allāh to indicate "disputes. . . corruption and much shedding of blood"; this probably refers to the chain of events between the murder of 'Uthmān on 17 June 656 and that of 'Alī on 24 January 661, including the famous arbitration between 'Alī and Mu'āwiyah which took place at Adhruḥ in January 659. The "fighting and wars" include the "Battle of the Camel" near al-Baṣra on 9 December 656 and that fought at Ṣiffīn near al-Rāqqa on 26 July 657. Māshā'allāh says this deplorable state of affairs continued until Saturn entered Scorpio; this, of course, occurred in 660 rather than 661.

Yazdijird must die in the East because the Sun, which indicates his demise, is in the ascendant; we have already seen that he was murdered near Marw. The conjunction taking place in the triplicity of water in the hypogee indicates the spoiling of the crops from water; see *Epistola*, chapter 10: "In aquaticis vero significat damna pluviarum ac pestilentiae, ea tamen conditione si sint impediti." This lasts for two years, since it will take that long for the two planets to enter Leo.

The shift from the triplicity of air to that of water had, of course, taken place in 571; and there is no reason to choose Pisces as the significant sign of the triplicity of water. Still, Pisces does indicate Ṭabaristān. Christensen (pp. 508-509), following Mas'ūdī, lists two sons and three daughters of Yazdijird III. Of the sons, Bahrām's fate is unknown, but Fīrūz died in China in 672; of the daughters, the destinies of Adragh and Mardāvand remain obscure, but there is a Shī'a tradition that Shahrbānū married Ḥusayn, the son of 'Alī. Still, we know that a certain Kisrā who allied himself with the Turks in 728/729 was a descendant of Yazdijird; so some remnants of the family did survive. Mas'ūdī says that they stayed at Marw; but that some took refuge in Ṭabaristān is not at all impossible. Before Yazdijird's flight to Seistan in the mid 640's the spāhbadh of Ṭabaristān had offered him an asylum and these spāhbadhs maintained their independence against the Muslim invaders for over a century. This was a natural refuge for an heir to the throne of Sāsān. Abū Shujā' Buwayh, the father of the man who founded the Buwayhid dynasty in 946, claimed descent from the Sasanian royal line; and he came from Ṭabaristān. Therefore, it is possible that Māshā'allāh is referring to a real survival of Yazdijird III's descendants in Ṭabaristān; but it is more likely that Ibn Hibintā (who wrote in about 950) has inserted this astrologically unjustifiable passage to support the dynastic pretensions of the then rulers of the caliphate.

14-16. Jupiter, in the third place cadent from the hypogee and in conjunction with Mars, is badly injured; and Jupiter indicates 'Irāq (see *De revolutione annorum mundi*, chapter 3). The Sun is injured through quartile aspect by Saturn; and the Sun also indicates 'Irāq. And finally, the moon is injured by conjunction with its descending node; its house, Cancer, indicates the fourth climate. Further, the significator regis is either the Sun or the lord of the tenth place; the beginning of the tenth place is in Sagittarius by division, and Jupiter is its lord. Thereby these considerations confirm the predictions previously made concerning Yazdijird III.

17. Mean conjunction 325 occurred on 18 March 670.

18. Mean conjunction 326 occurred on 22 January 690.

f216r: 1. Mean conjunction 327 occurred on 30 November 710.

2. Mean conjunction 328 occurred on 5 October 729.

2-7. The interval between mean conjunctions 323 and 329 is (19^y 10^m 11^d) X 6 = 119^y 2^m 6^d. Since we know from 231r: 19 that mean conjunction 323 occurred 1^y 9^m 17^d before Yazdijird, it follows that mean conjunction 329 occurred 117^y 4^m 19^d after Yazdijird. Ibn Hibintā wrongly reads 6^m (see 217v:9-10), but a glossator corrects it. As usual, Ibn Hibintā's calculation will not work; these are not real years that he is working with. The glossator further states that mean conjunction 329 fell in Rabī' I of A.H. 132 or between 18 October and 17 November A.D. 749; in fact it fell on 11 August 749.

7-10. 17 Bahman 119 (read 117) Yazdijird equals 31 March 749; the glossator's date, 6 Sha'bān A.H. 131 also equals 31 March 749. Originally the date must have been given as 7 Bahman 117 Yazidijird = 21 March 749. The date of the horoscope is 19 March 749.

f216v. Horoscope 9

	Text	Computation
♄	♏ 19 retr.	♏ 23 retr.
♃	♏ 6;8 retr.	♏ 11 retr.
♂	♓ 19	♓ 22
☉	♈ 0;1	♈ 2
♀	♉ 17;6	♉ 18
☿	♓ 27;12 retr.	♓ 19 retr.
☾	♒ 23;5	♒ 14
H	♏ 25	
☊	—	

1. The lord of the year travels at the rate of one zodiacal

sign per year; therefore, beginning from Libra, the ascendant of the horoscope of the vernal equinox of 571, it reached Leo. The division, traveling at the rate of 1° per year, was said (rather erroneously) to have been in Pisces 20° in 570; this puts it in Virgo 19° in 749. It is aspected diametrically by Mars and by Mercury in retrogression, both of which bode ill.

f217r:2-4. The lord of the year is Mars because:
(a) It is lord of the ascendant, Scorpio;
(b) It is aspected (1) by Saturn and Jupiter in quartile,
(2) by Venus in sextile.

The fifth place by counting is, of course, Pisces; and they are aspected in quartile (bad), not trine (good) by Saturn. Mars "receives" Saturn in his house, Scorpio. Though Pisces is not a human sign, the prediction of pestilence at the beginning of the year can be in part explained from *De revolutione*, chapter 221: "Scito quod cum fuerint mali Saturnus scilicet et Mars in signo ad imaginem hominum et iunctus fuerit et alter malus – fueritque directus aut retrogradus – ex coniunctione ac quarto aspectu vel oppositione, erunt in hominibus pestilentiae; si vero fuerit retrogradus, erit calidius atque velocius."

5-7. Venus from Taurus, its own house, aspects Saturn diametrically (injuriously) in Scorpio, the ascendant, which indicates the East; and Saturn, itself in retrogression, is in conjunction with Jupiter, also in retrogression. The predictions result from the following rules: *De revolutione*, chapter 36: "Et cum fuerit Saturnus (dominus anni!) in ascendente, significat famem, mortem, infirmitates ac ventos vehementes ac noxios"; *ibid.*, chapter 46: "Et si fuerit Saturnus dominus anni (!) et fuerit in aquaticis signis in angulo, grave erit frigus et multiplicabuntur locustae; et si fuerit retrogradus, erit discordia maxima et mors valida." See also *Epistola*, chapter 10: "Si vero Saturnus praeerit (in a coniunctio maior) significabit detrimentum atque tribulationem;" Saturn is stronger than Jupiter because it is so near its exaltation. One may also cite *De revolutione*, chapter 15: "Scito quod planeta tardior significat bellum, et retrogradus significat fugam."

7-10. The lord of the ascendant, Scorpio, is Mars, which aspects Saturn in trine from the fifth place; the lord of the ninth place, Cancer, is the Moon, which aspects Saturn in quartile from the hypogee; and the lord of the fourth place,

Aquarius, is Jupiter, which is in conjunction with Saturn. Further, Saturn is in the ascendant in the upper hemisphere. Therefore the predictions of *De revolutione*, chapter 37, should hold: "Scito quod esse anni gravabitur cum fuerint mali in parte circuli superioris et in parte septentrionali, et maxime si commiserit Mars dispositionem suam Saturno; quia Saturnus significat res graves et diuturnas, de apparitione detrimenti in terra, et debilitatem regis et ablationem fidei ac religionis; et multiplicabuntur latrocinia et contentiones, et maxime si fuerit aliquis malorum in medio coeli. Et aspice cum hoc Lunam, quia si commiserit dispositionem suam Saturno ex inferiori parte circuli subterranei significat tribulationem ac detrimentum et mutationem regni. Et si Luna pulsaverit lumen suum Saturno et fuerit cum eo aut in quarto aspectu vel in oppositione, erit tanto deterius." Instead we are told that the people of the East (= the ascendant) will rule.

10-12. However, the evil things do affect the Arabs, whose sign Scorpio is and whose planet, Venus, though in its own house, is injured by diametric aspect from Saturn and sextile aspect from Mars.

12-18. Mars, the lord of the ascendant and the lord of the year, indicates war; see *De revolutione*, chapter 21: "Post hoc aspice bellum ex parte qua fuerit Mars." Pisces, the sign it occupies, is one of two bodies; and presumably the war is in the fifth year because Mars is in the fifth place. But I do not understand the connection between these two statements. However, Mars' presence in a sign belonging to the triplicity of water probably explains why the war will occur in the western region; Cancer, Scorpio, and Pisces govern the West. However, it is the people of the East who get killed because Mars is the lord of the ascendant. Mercury, being retrograde and in conjunction with Mars, indicates that unpleasant things will befall the people indicated by Mars' house, Scorpio – that is, the Arabs.

f217v:1-3. By division, the tenth place begins in Virgo, which is Mercury's house; so Mercury, which is retrograde and in conjunction with Mars, the lord of the year, aspecting the ascendant, is the significator regis. Therefore, we find this prediction in Chapter 19 of the *De revolutione*: "Post hoc aspice cui coniungatur significator regis; quia cum coniungitur domino anni, committet dispositionem suam rusticis; et si separatus fuerit a domino anni (!) et iunctus fuerit malo planetae qui ascendenti inimicus fit, supplicabitur super eos et rex immittet super eos impedimentum."

The events posteriorly predicted by Māshā'allāh began before 19 March 749; for Abū Muslim raised the 'Abbasid black banner of revolt against the Umayyid caliphs (the "Westerners") on 9 June 747, and, at the head of an army of Iranian ("Eastern") peasants, soon took Marw, the capital of Khurāsān. However, it was not till 30 October 749 that Abū al-'Abbās was enthroned as caliph in al-Kūfah. Damascus was taken on 26 April 750, and the last remnants of the Umayyid house were massacred at a banquet on 25 June 750. Thus the "people of the East" gained ascendancy over the Arabs.

The battle in which "Easterners" fought in the West could possibly be 'Abdallāh's pursuit with his Khurāsānian army of the shattered forces of Marwān through Syria, Palestine, and Egypt in the first half of 750. But the statement that it occurred in the fifth year and that many "Easterners" perished makes more likely Abū Muslim's defeat of 'Abdallāh in November 754 at Nisibis; presumably both armies were primarily composed of Iranians.

Al-Manṣūr's suppression of the Shī'a revolt led by Ibrāhīm and Muḥammad in 762-763 was perhaps sufficient cause to speak of "death and killing and brutality and oppression on the part of the Sulṭān"; but Māshā'allāh is more likely referring to his crushing of the revolt led by Sunbād the Magus in Khurāsān in 755.

6. Mean conjunction 330 fell on 18 June 769.

7. Mean conjunction 331 fell on 25 April 789.

8. $(19^y\ 10^m\ 11^d) \times 3 = 59^y\ 7^m\ 3^d$.

9-10. From 216r:5 we have the erroneous statement that from the beginning of the Era of Yazdijird to mean conjunction 329 was $117^y\ 6^m\ 19^d$; if we add $59^y\ 7^m\ 3^d$ we get that the interval between the beginning of the Era of Yazdijird and mean conjunction 332 was $177^y\ 1^m\ 22^d$. The glossator places the mean conjunction in Sha'bān A.H. 193, or 20 May – 18 June A.D. 809; in fact it fell on 2 March 809.

12-13. 22 Bahman 177 Yazdijird = 20 March 809; the horoscope is dated 20 March 809.

14-16. From the ascendant of the vernal equinox of 571, Libra, the year-sign has reached Cancer in 809; from the sign of conjunction, Scorpio, it has reached Leo. The division has reached Scorpio 19°.

f218r. Horoscope 10

	Text	Computation
♄	♐ —	♐ 6 retr.
♃	♏ 29;32	♐ 5 retr.
♂	♒ 23;45	♒ 29
☉	♈ 0;1	♈ 4
♀	♓ 16;10	♓ 18
☿	♓ 19	♓ 7
☾	♈ 16;39	♈ 12
H	♋ 3	
☊	♋ 28;18	

1-5. Venus is lord of the year because:

(a) It is diurnal lord of the triplicity of water, to which Cancer belongs; and this is a diurnal horoscope;

(b) It is in (the sign of) its exaltation, Pisces;

(c) It is in the tenth place by division;

(d) It is "received" in his house (Pisces) by Jupiter, whose exaltation is in Cancer (15°), the ascendant;

(e) It is aspected in trine by Jupiter, the lord of Pisces.

5-6. The nature of Cancer is cold and wet (water) while that of Venus is hot and wet (air). Perhaps all that is referred to is Venus' diurnal lordship of Cancer's triplicity and its trine aspect of the ascendant. The six years are derived from the six signs between Mars, the lord of the tenth place by counting, and the ascendant.

6-218v:1. "The people of their kingdom" are the people of 'Irāq, for Cancer indicates 'Irāq. The "people of the East" join in because Cancer is the ascendant, and those of Iṣfahān because the conjunction of Jupiter and Saturn occurs in Sagittarius, which indicates that city.

f218v:1-2. The four years are predicted from the four signs between the significator regis, the Sun, and the ascendant.

2-4. Death increases in the tenth year of the conjunction and the government will be transferred to the East because then Saturn will be in conjunction with the Sun, the significator regis, and because the conjunctions have shifted from the triplicity of water to that of fire, which indicates the East.

4. I do not know why the years of the kings are lengthened; see 219r:13-15.

4-7. The people will be killed because Mars, the lord of the eleventh place (Aries) by division, is in the eighth place (Aquarius) by counting, and the eighth place is the place of death; moreover, it is aspecting in sextile (injuriously) Saturn, Jupiter, the Sun, and the Moon. The evil thus generated is mitigated by Saturn's reception of Mars in its house, Aquarius, with sextile aspect. The identification of these events with real historical occurrences is fraught with difficulty. We know that Māshā'allāh died in about 815, in the sixth year of the conjunction; how much earlier he wrote the *Kitāb fī al-qirānāt wa al-adyān wa al-milal* is not known. The reference to the transfer of the rulership from one house to another in the fourth year may be taken as indicating the enthronement of al-Ma'mūn as Caliph in 813; but his being a brother of his predecessor, al-Amīn, makes this doubtful. In any case, it is clear that Māshā'allāh expected the Caliphate to be overthrown in the sixth year (815) by a coalition of 'Irāqians and Iranians, with the rulership passing to an Iranian. Māshā'allāh, from this passage and from his previous attack on the 'Abbāsids (217v:3), can be judged a partisan of Iranian nationalism. Bīrūnī[12] says that Kanaka, the famous Indian astrologer-astronomer of the time of al-Manṣūr and Hārūn al-Rashīd who wrote two books on Saturn-Jupiter conjunctions, determined the time of the fall of the 'Abbāsid dynasty, and that its successor would come from Iṣfahān.

7-219r:5. In this long passage Ibn Hibintā justifies the failure of the shift from one triplicity to another to indicate the loss of the kingdom by the Arabs. First he compares Horoscopes 6 and 10, and finds that Cancer, the ascendant in the latter, was the tenth place in the former (Cancer, of course, is also the moving sign in the triplicity of water to which Scorpio also belongs).

12. *The Chronology of Ancient Nations*, trans. C. E. Sachau (London, 1879), p. 129.

The Moon, which is the lord of Cancer and, as significator regis, determined the career of the Prophet in the interpretation of Horoscope 6, was then in Cancer in the tenth place by counting. In Horoscope 10 it is again in the tenth place by counting; furthermore, it is waxing and is received in its house by Mars, which aspects it in sextile. The similarities between the positions of Cancer and the Moon in the two horoscopes, especially since Cancer indicates 'Irāq and the Moon the Prophet, assure the continuity of the Arab state. However, the shift to the triplicity indicating the East indicates victories for the Khurāsānians and Turks during the shift, but without the Arabs' actually losing their dominion until the next shift, to the triplicity of earth, in 1047. It is interesting to note that the Saljūq Turk Ṭughril Beg entered Baghdād and put an end to the Buwayhid dynasty in 1055, about a century after Ibn Hibintā wrote.

5-13. While the conjunctions were in the triplicity of air between 333 and 571 (mean conjunctions 308 to 319), the Sasanians ruled (astrological theory thus cheats them of a century of their reign); the shift to a new triplicity, however, in 571, indicated the transfer of the rule to the Arabs. There followed a period of instability until the fourth conjunction of the triplicity in 630 – which, like the first, occurred in Scorpio – finally resolved the issue definitively. The point is that the Sasanian dynasty could not survive because there were allegedly no connections between the horoscopes of the vernal equinoxes of 333 and 571 as there are between those of 571 and 809. For Māshā'allāh's horoscope of the year-transfer for conjunction 308, see above, p. 82.

15-19. Here Ibn Hibintā inserts a note concerning the troubles of his own times, during the last years of the seventh conjunction (928-948) (see 224v:3 - 225v:19) when Saturn conjoined with Mars in Cancer (in 945). The conjunction of 928 was in Sagittarius; but also the year-sign had proceeded from Libra in 571 to Sagittarius in 945. The king changes after 15 years because Mars, the significator regis, is 15 signs from the ascendant; this refers to al-Mustakfī's succession to al-Muttaqī in 944. The conjunction of Saturn and Mars in Cancer in 945 foreboded even more terrible events; see, for example, Abū Ma'shar, *Kitāb al-qirānāt* and *Kitāb iqtirān al-naḥsayn fī burj al-Saraṭān*. But Ibn Hibintā is careful not to mention that this year saw the establishment of the

Buwayhid dynasty; he substitutes the turmoil predicted from the eighteenth year (945) on. The first of the dynasty, Aḥmad ibn Buwayh, was succeeded in the second year of the eighth conjunction (949) by the greatest of the line, ʻAḍud al-Dawlah, under whom Ibn Hibintā probably wrote.

f219v: 1. Mean conjunction 331 fell on 6 January 829.

2. 197 Yazdijird extended from 28 April 828 to 27 April 829.

3-4. Horoscope 11 is dated 20 March 829.

	Text	Computation
♄	♌ 6	♌ 5 retr.
♃	♋ 25;44	♌ 2
♂	♌ 2;7	♌ 26
☉	♈ 0;1	♈ 4
♀	♉ 15;30	♉ 19
☿	♓ 4	♓ 6
☾	♌ 28;18	♌ 22
H	♌ 30	
☊	♋ 1	

f220r: 1-2. Mars and Saturn, while both in the ascendant sign Leo, are cadent in the twelfth place by division.

2. The Moon in the ascendant aspects Mercury diametrically.

2-3. The Sun in the ninth place by counting aspects Saturn in trine (its aspect of Mars does not count because Mars receives the Sun).

3-4. Jupiter in its sign of exaltation (Cancer 15°) is in the eleventh place by division, cadent from the twelfth.

4. Mercury is in the eighth place by counting, aspected diametrically by the Moon.

5-6. Venus, retrograding, in its own house, is in the tenth place by counting, aspecting Mars in quartile, injuriously; but why "against it in heaven and house"? Venus, being in its own house, is not received, as the Sun is.

6-7. Mercury is indeed in the sign of its dejection (Pisces 15°); it is also in the tenth and seventh place, respectively, from its two houses, Gemini and Virgo.

The lord of the year is the Sun, because:

(a) It is lord of the ascendant, Leo;

(b) It is aspected by (1) Saturn, Mars, and the Moon in trine, (2) Jupiter in quartile;

(c) It is received by Mars in its house, Aries;

(d) It is in the sign of its exaltation (Aries 19°).

The significator regis is Venus, the lord of the tenth place, which is in its own sign, Taurus, but is aspected in quartile (injuriously) by Mars, Saturn, and the Moon and in sextile by Mercury.

7-220v: 1. All these predictions are, of course, due to the imagination of Māshā'allāh. The king's opponent is determined by Mars, which aspects the significator regis, Venus; Mars' being in Leo, which belongs to the triplicity of the East and is the ascendant, indicates an eastern origin for the opponent. The degeneration of the general populace is indicated by *De revolutione*, chapter 15: "Post hoc aspice malum qui aspicit dominum anni aut significatorem regis; quia si aspexerit ab ascendente, significat quod ipsum impedimentum sit in communi." The weakness of the king is shown by the retrogression of Venus; *ibid.* chapter 20: "Retrogradatio significatoris regis significat debilitatem regis." The corruption of religion is indicated by the Sun's being aspected by Saturn in the ninth place. All this occurs in the first year because Mars is in the ascendant.

There is fighting and bloodshed in the second year because Mercury, injured, is in the eighth place in a sign of two bodies; the third year brings grief owing to the three signs between Venus and the ascendant. The disasters of the eighth and ninth years are determined respectively by the number of signs between Mars and Mercury and between Mars and the Sun.

f220v: 1. Mean conjunction 334 fell on 12 November 848.

1-2. 216 Yazdijird equals 23 April 848 – 22 April 849.

3-4. Horoscope 12 is dated 20 March 848.

	Text	Calculation
♄	♓⟨1⟩5;55	♓ 22
♃	♓ 11	♓ 17

♂	♏ 8;12 retr	♏ 17
☉	♈ 0;1	♈ 4
♀	♉ 10;12	♉ 15
☿	♓ 11;22	♓ 12
☾	♍ 5;43	♍ 1
H	♌ 4;28	
☊	♊ 23	

5-6. The Sun is lord of the year because:
(a) It is lord of the ascendant, Leo;
(b) The horoscope is diurnal.
6-7. The Sun is not aspected by either Mars or Saturn, and therefore this is a good indication for its climate, the fourth (including Baghdād).

f221r: 1-7. The disasters in the East are caused by Mars in retrogression, aspecting the ascendant, Leo, in quartile, injuriously. Mars' opposition to Venus, again the significator regis and, though in its own house, also aspected by Saturn, by Mercury in its dejection (and by Jupiter) in sextile (and by the Moon in trine), should cause the ruler of the Arabs some trouble; and that whole nation must suffer because Mars is retrograde in the hypogee in their sign, Scorpio, aspected by Saturn, by Mercury (and by Jupiter) in trine. This occurs in the first year because Scorpio is a fixed sign, in the fifth because of the number of signs between Mars and Saturn. The misfortunes of Persia and Andalusia arise respectively from Mars' diametrical and Saturn's and Mercury's sextile aspect of Taurus and from Saturn's and Mercury's diametrical and Mars' sextile aspect of Virgo. The violence followed by success in Ṭabaristān results from Saturn's and Mercury's being in Pisces aspected in trine by Mars, while Jupiter is there also aspected in sextile by Venus and diametrically by the Moon.
7. Mean conjunction 335 fell on 16 September 868.
8-9. 236 Yazdijird equals 18 April 868 – 17 April 869.
9-10. This horoscope is dated 20 March 868.

f221v. Horoscope 13

	Text	Computation
♄	♐ 0;18	♐ 8 retr.
♃	♏ 19;57 retr.	♏ 25 retr.
♂	♉ 17;34	♉ 28
☉	♈ 0;1	♈ 5
♀	♓ 2;5	♓ 6
☿	♓ 6;59	♓ 9
☾	♑ 13;12	♑ 11
H	♍ 9;39	
☊	♌ 8;9	

1. The cardines of the ascendant are perpendicular (more or less) because the ascendant degree is close to Libra 0°; the beginnings of the fourth and tenth places, then, will be approximately equidistant from it.

1-3. Saturn, in Sagittarius 0;18°, is indeed cadent to the fourth place by division; this doubly confirms our emendation. Venus and Mercury are cadent to the seventh place by division. The Sun, at Aries 0;1°, is indeed 2° short of the center of the seventh place by division, which is at Aries 1;48°; this confirms our assumption that Māshā'allāh used a system of rising-times leading to identical results with Babylonian System A. The Sun aspects Saturn in trine.

3-4. Mars is in the ninth place by counting, and the Moon in the fifth aspecting Jupiter in sextile. The Moon, in Capricorn, Saturn's house, is not received because Saturn does not aspect it.

The lord of the year is the Sun because:

(a) It is in a cardine, the seventh place, by division;

(b) It is aspected (1) by Saturn in trine, (2) by the Moon in quartile;

(c) The horoscope is diurnal;

(d) It is in the sign of its exaltation.

The significator regis is Mercury, the lord of the tenth place, Gemini.

4-6. Saturn, aspecting the Sun in trine and the tenth place in opposition, indicates that the king of the fourth climate will perish; the death of the king can also be inferred from Mercury's being in the sign of the dejection (Pisces 15°) aspected in quartile by Saturn and in sextile by Mars. That this unfortunate event will occur in the fourth year is known from the number of signs between Mars and the ascendant (cf. 222:4). Why a member of his house will not inherit the throne is unclear.

6. The butchery in Iṣfahān results from the conjunction's occurring in Sagittarius; see 218v:1.

7. For the rules governing the king's travel, see *De revolutione*, chapter 42.

7-222:1. The misfortune of the Arabs arises from Jupiter's retrogression in and Mars' diametric aspect of Scorpio.

f222r:1-6. The sixteenth year is bad because there are sixteen signs between Mars and the ascendant. The second climate is indicated by Jupiter; religion by Mars' being in the ninth place; and Mawṣil by an unidentifiable factor.

7-8. The crawling animals and cattle are signified by Scorpio and Taurus; the goats, under Capricorn, are safe (as are the sheep under Aries and the lions under Leo).

8-11. Saturn, governing the first climate, is not injured by Mars; Jupiter and the second are injured; Mars and the third are the worst; the Sun and the fourth are not injured by Mars; Venus and Mercury, indicating the fifth and the sixth, are aspected in sextile by Mars; and the Moon, ruling the seventh, is aspected in trine by Mars.

12. Mean conjunction 336 fell on 24 July 888. 257 Yazdijird equals 13 April 888 – 12 April 889.

14. This horoscope is dated 20 March 888.

f222v. Horoscope 14

	Text	Computation
♄	♌ 2;56	♌ 7 retr.
♃	♋ 18;20	♋ 23
♂	♒ 18;27	♒ 29
☉	♈ 0;1	♈ 5
♀	♉ 13;32	♉ 15
☿	♓ 4;27	♓ 7
☾	♉ 22;0	♉ 22
H	♏ 18	
☊	♈ 28	

1-4. Mars is aspected:
(a) By Saturn in opposition;
(b) By the Sun in sextile;
(c) By Venus in quartile;
(d) By the Moon in quartile.
Venus is:
(a) In its own house, Taurus, in a cardine, the seventh place;
(b) Aspected by (1) Saturn in quartile, (2) Jupiter in sextile, (3) Mars in quartile, (4) Mercury in sextile.

Venus aspects Jupiter in sextile, Jupiter being in the sign of its exaltation; I do not know why it does not "receive" the aspect. In fact, Jupiter is received in its house by the Moon, which also aspects it in sextile. But we are told that the Moon, the nocturnal luminary (it is a nocturnal horoscope), aspects Saturn in quartile, and Saturn accepts this aspect; in fact, Saturn is received in its house by the Sun, which aspects it in trine.

The lord of the year is Mars because:
(a) It is lord of the ascendant, Scorpio;
(b) It enjoys the many aspects enumerated above;
(c) It is in a cardine, the fourth place.
The significator regis is the Sun, lord of the tenth place, Leo.

4-223r:9. Saturn, in the tenth place, aspects the Sun, the significator regis, in trine, and the Moon (also significant of royalty in nocturnal horoscopes) in quartile; thence hardship for kings, especially in Khurāsān, which is indicated by Leo. The seventh climate is affected because of the injury to the Moon; the enemy appears in the fourth climate because of the injury to the Sun. There are ten signs between Saturn and the Moon, and four between Saturn and the ascendant, indicating the respective years. Jupiter, Venus, and Mercury are not free of aspects from Mars, Saturn, or both; and, though the first is in the sign of its exaltation and the second in its own house, Mercury is in the sign of its dejection. It is, therefore, difficult to see why Māshā'allāh reaches such an optimistic conclusion concerning them.

f223r:9. Mean conjunction 337 fell on 27 May 908.

10. 27 May 908 is the 20th day of the second month of 279 Yazdijird.

12. This horoscope is dated 21 March 907.

f223v. Horoscope 15

	Text	Computation
♄	♓15;28	♓24
♃	♓4;34	♓9
♂	♓20;45	♓27
☉	♈0;1	♈5
♀	♓22;37	♉4
☿	♓15;45 retr.	♓6 retr.
☾	♊8;22	♉29
H	♐23	
☊	—	

1-224r:4. ‘Abdallāh ibn Muḥammad ibn Bishr is otherwise unknown. He is correct in asserting that Māshā’allāh should have added the equation instead of subtracting it; however, his other statements regarding Venus are wrong. The equation should, of course, be added to the mean Sun, which is in Pisces 27;49°, rather than to Aries 0°. This would put Venus (minus the equation of the center) correctly at Taurus 3;7,34°. Māshā’allāh, in erroneously subtracting 35;23° from Pisces 27;49°, should have arrived at Aquarius 22;26°.

The ascendant of Horoscope 14 was Scorpio 18°; 19 X 93;15° = 331;45°, which would put the ascendant of Horoscope 15 in Libra 20°. Māshā’allāh should have subtracted 28° from Scorpio 18°, but added them instead to put the ascendent in Sagittarius. The parameter for the excess of revolution, 93;15°, corresponds to a year-length of 6,5;15,32,30^{d}, which is attested elsewhere for the *Zīj al-Shāh*. It is indeed the year-length used by Māshā’allāh.

5-6. The five planets in the fourth place by counting, Pisces, are Saturn, Jupiter, Mars, Venus, and Mercury; but we have seen that Pisces is really the sixth place, and Venus is really in Taurus.

6-7. The Moon in Gemini, the seventh place by counting, aspects Saturn (and the other planets in Pisces) in quartile.

7-8. The Sun in Aries is in the fifth place by counting but the fourth by division; as being in the fifth, it is cadent. It is not received by Mars in its house, Aries, because Mars does not aspect it.

8. Venus and Mars, being within 15° of the Sun, are under its rays.

8-9. Jupiter, the lord of Sagittarius, the ascendant, is in its own house, Pisces, in the eastern hemisphere.

The lord of the year is Jupiter because:

(a) It is lord of the ascendant, Sagittarius;

(b) It is aspected by the Moon in quartile;

(c) It is in its own house, Pisces;

(d) It is a masculine planet in the eastern hemisphere.

The significator regis is Mercury, the lord of the tenth place, Virgo.

9-224v:3. The predictions are again the product of Māshā'allāh's imagination. Aquatic animals are affected because the lord of the year is in conjunction with Saturn and Mars in an aquatic sign, Pisces; this also explains the drowning. The blindness arises from the ascendant, Sagittarius.

The South and the West are injured through the two houses of the significator regis, Mercury, which is retrograde in the hypogee and in its dejection in conjunction with Saturn and Mars. Mercury aspects Virgo, which belongs to the triplicity indicating the South, diametrically, and Gemini, which belongs to that indicating the West, in quartile, both injurious aspects. The people of the West are afflicted in the seventh year because Gemini is the seventh place from the ascendant; in the sixteenth because there are sixteen signs between Mars and Gemini. Violence strikes the fifth and third climates because of the unfortunate positions respectively of Venus and Mars.

Kings die in the fourth year owing to Mercury's being with Mars in the fourth place from the ascendant. The good accruing to the people of prophecy and science can only be explained by the fact that the place of religion, the ninth, Leo, is aspected in trine by its lord, the Sun, and in sextile by the Moon, while the Moon is received in its house, Gemini, by Mercury with quartile aspect. And Leo is not aspected by Saturn or Mars.

f224v:3-4. Mean conjunction 338 fell on 3 April 928; 1 Farvardīn 297 Yazdijird equals 3 April 928 (see also 225r:19–225v:1).

5. Horoscope 16 is dated 21 March 928

	Text	Computation
♄	♐ 12;1	♐ 22

♃	♐12;11	♐19
♂	♈28;13	♉ 6
☉	♈ 0;1	♈ 6
♀	♓7;46	♉20
☿	♓19;14	♓10
☾	♒29;22	♒28
H	♋8	
☊	—	

f225r:1-3. The Sun is lord of the light in the diurnal horoscope and is in the tenth place by counting, the eleventh by division. It aspects Jupiter in trine.

3-4. Jupiter and Saturn conjoin in the sixth place, Jupiter receiving Saturn in its house, Sagittarius.

4-6. The Moon, the lord of the ascendant, is in a cadent, the eighth place, and is not increasing in motion; it is also waning. It aspects Jupiter in sextile.

6. Mars is in the tenth place, Aries, by counting.

6-7. Venus, in its own house, Taurus, is in the eleventh place by counting, not aspected by either Saturn or Mars.

7-8. Mercury in Pisces is in its dejection in the ninth place by counting, the tenth by division.

The lord of the year is the Sun because:

(a) It is in a cardine, the tenth, in a diurnal horoscope;

(b) It is received by Mars in its house, Aries, through conjunction;

(c) It is aspected by Saturn and Jupiter in trine and by the Moon in quartile;

(d) It is in the sign of its exaltation.

The significator regis is Mars, the lord of Aries, the tenth place.

8-13. Again this is all fanciful. Sickness comes about through the Sun's being in conjunction with Mars and aspected by Saturn; sorrow to the king because Mars is his signifier and is aspected also by Saturn. Babylon is affected through Cancer, the ascendant, which indicates 'Irāq, and the Sun, which indicates the fourth climate. The six years correspond to the number of signs between the ascendant and Saturn; the eighteenth year is simply six plus twelve. Venus and the Moon indicate good respectively for their people because neither is aspected by Mars.

14-17. Ibn Hibintā's criticism is not justified, since each set of predictions can be justified from the horoscope.

18-225v:19. Ibn Hibintā, in order to obtain a horoscope which will produce predictions conforming to the events leading up to the founding of the Buwayhid dynasty (and thereby justifying it), now breaks the rules of the revolutions of the world-years and casts the horoscope of the actual conjunction of 928. He does not, however, use the *Zīj al-Shāh* as had Māshā'allāh, but rather the *Zīj al-Mumtaḥan* of Yaḥyā ibn Abī Manṣūr.

f225v:2-3. 9 Mihr 297 Yazdijird equals 8 October 928.

2-4. Horoscope 17

	Text	Computation
♄	♐16;41	♐17
♃	♐16;41	♐15
♂	♍16;25	♍17
☉	♎21;21	♎20

5-7. Mars is in the tenth sign from Saturn and Jupiter. For the astrological reasons for Mars' and Saturn's superiority, see Bīrūnī's *On transits.* We cannot reconstruct Ibn Hibintā's calculations, since we do not possess a copy of the *Zīj al-Mumtaḥan.*

7-12. The Caliph al-Muqtadir was briefly deposed in the eleventh month (Dhū al-Qa'da 316 equals December 928 – January 929) by the chief of the bodyguard, Mu'nis al-Muẓaffar, in favor of his half-brother al-Qāhir; however, he soon regained the throne and continued as Caliph till his murder by Berber soldiers in the eighth month of the fifth year of the conjunction (Sha'bān 320 equals August – September 932), when he was again succeeded by al-Qāhir. For the murders of Nāzūk and Abū al-Hayjā, see al-Hamadānī, *Takmila ta'rīkh al-Ṭabarī* (Imprimerie Catholique, Beirut, 1959), vol. 1, pp. 77-78.

13-14. Al-Qāhir was deposed and his eyes were put out in the second month of the seventh year of the conjunction (Ṣafar 322 equals January-February 934); al-Rāḍī succeeded him on 24 April 934.

15-17. The Sājiya were a body of troops raised by and named after Ibn Abī al-Sāj.[13]

17-19. In the thirteenth year of the conjunction, 940, al-Rāḍī died and was replaced by al-Muttaqī.

19. The turmoil presumably ended, in Ibn Hibintā's sycophantic version, with the entry of Aḥmad ibn Buwayh into Baghdād and his deposing of al-Mustakfī and installation of al-Muṭīʿ in the seventh month of the eighteenth year of the conjunction (Rajab 333 equals January 945). For his less favorable interpretation of Aḥmad ibn Buwayh's action, carefully concealed in the discussion of Horoscope 10 (and deliberately separated from the present account by pages of meaningless conjecture on Māshā'allāh's part), see 219r:15-19.

13. See, for example, H. Bowen, *The Life and Times of ʿAlī ibn ʿĪsā ʿThe Good Vizier'* (Cambridge University Press, Cambridge, 1928), p. 327.

APPENDIX 1

Māshā'allāh's Date of the Prophet's Birth

This date is recorded in his *Chronology*[1] by Bīrūnī as being Monday, the day Khūr (11) of the month Dai in the year 41 of Anūshirwān at the beginning of the seventh hour with the Sun in "the liver of heaven" and the ascendant in Cancer. 1 Favardīn of the 41st year of Anūshirwān according to Māshā'allāh's calendar was 3 May 571; and therefore 11 Dai, the day of the Prophet's birth, was 7 February 572, which was indeed a Monday. The Sun was then in Aquarius, and would have been setting if Cancer had been in the ascendant. This passage in Bīrūnī confirms our previous assertions regarding the *zīj* used by Māshā'allāh and its calendar.

1. *Documenta Islamica inedita,* ed. J. Fück (Akademie Verlag, Berlin, 1952), pp. 95-96.

APPENDIX 2

Māshā'allāh's *Fī qiyām al-khulafā'*
wa ma'rifa qiyām kull malik

(This translation is based on two late manuscripts, B = Berlin Ar. 5898, ff. 46r-47v, and V = Vatican Ar. 955, ff. 249r-250v. Important variants are given in the footnotes.)

In the name of God, the Merciful, the Compassionate. [1]This is[1] the book of Māshā'allāh the astrologer on the accession of the Caliphs and the knowledge of the accession of every king. He said: I have described to you elsewhere about kings, by analogy and directly; [2]it is just[2] as [3]I told you[3] at these places [4]the secrets[4] (?) [5]in connection with these[5] [6]years[6] and [5]with this[5] period of time, if God, He is exalted, wishes.

Know that the Sun is the regent of the stars and the dispenser of time. Through it do the planets become eastern and western, have [7]forward[7] and retrograde motion; through it are known their turnings, mansions, and conditions. All the planets are attached to the Sun as are subjects to their rulers. The Sun is the indicator in the matter of kings to the exclusion of these (the planets), except that Saturn, Jupiter, or Mars may have partnership with it; this (share) is a seventh. But the Sun[8] takes away the seventh if they are at its eye (?) or its left (north) or under its control, if God,

1-1 om. V. 2-2 year (*sina*):BV. 3-3 you want:V. 4-4 om. B. 5-5 for a king:B. 6-6 motion:B. 7-7 om. V. 8 in the middle:add. V.

He is exalted, [9]wills[9]. The Sun [10]among them[10] [11]is like[11] a monarch in the middle of his kingdom.

Know the year ⟨in⟩ which [12]the accession[12] occurred.

If a ruler accedes to power, know the year in which he accedes and know what [13]I told you[13] concerning kings. The matter of kings is known from the Sun, Saturn, Jupiter, and Mars; the Sun gives the years (of the reign), and these [14]planets[14] sometimes increase, sometimes decrease them. I will clarify [15]this[15] in the [16]commentary[16] of the *Kitāb al-ma'ānī*[17] if God, He is exalted, wishes.

If you wish to know this, calculate the longitudes of [18]the planets[18] at the entrance of the Sun into the first minute of Aries; and know the ruler and the accession ⟨from⟩ these [19]three[19] superior and heavy [20]planets[20] – understand that Venus and Mercury do not belong with them – at the entrance of the Sun into the first minute of Aries.

If you find Saturn in the tenth ⟨place⟩ from the Sun or the eleventh, the Sun gives to the king according to the amount of its small years[21]. If there are 15° between Saturn and the Sun, then it gives a fourth of its small years; and as many[22] degrees as you [23]add, so many years add[23] until 30° is reached. If there are 65° ⟨between⟩ Saturn and the Sun, it gives a fourth of its small years; and as many[22] degrees as you subtract, they do not pass beyond this. If Saturn is in the tenth or eleventh ⟨place⟩ from it, the Sun gives a period of rule until Saturn and Jupiter aspect it from one degree and one sign or ⟨till⟩ they conjoin without aspecting.

If Saturn is in the seventh ⟨place⟩ from the Sun in the first degree [24]or first decan[24] of the sign, then it (the Sun) gives half its small years. If Saturn is [25]in the eighth[25] ⟨place⟩ from the Sun, and Jupiter has no partnership with it (Saturn), then it (the Sun) gives a fourth of its small years. If Saturn is in the eighth ⟨place⟩ from the Sun, and Jupiter has no partner-

9-9 commands:V. 10-10 in its house:B. 11-11 year:V. 12-12 the speech:BV. 13-13 is established for you:B. 14-14 om. B. 15-15 for you:V. 16-16 course:B. 17 Listed as a work of Māshā'allāh by Ibn al-Nadīm; fānā:add B, bābābābā: add. V. 18-18 om. V. 19-19 om. V. 20-20 om. B. 21 Two words in B (arranged equally?) and one in V (*qāmā*) add nothing to the sense. 22-22 om. V. 23-23 subtract, so many years subtract:B. 24-24 om. V. 25-25 in trine of the eighth.

ship with it (Saturn), and Mars is in the tenth ⟨place⟩, then it (the Sun) gives a seventh of its small years. If Saturn[26] is in right trine of the Sun and has traversed 20° of the [27]sign[27], it (the Sun) gives ⟨all⟩ its small years; [28]if Saturn is in right ⟨trine⟩, and Jupiter is in the eleventh ⟨place⟩ from it (the Sun), it (the Sun) gives three fifths of its small years.[28] If you see it (Saturn) in left trine, then sometimes it gives half of its small years.

[29]These[29] fore-mentioned examples are not set down for every king. For him, count from the sign ⟨occupied by⟩ the Sun – [30]one year for every sign[30] – till the place of Saturn or Mars; and according to this he dies, if God, [31]He is exalted[31], wills. [32]If he is safe from this place, ⟨count the signs⟩ till the other evil planet; but if God wishes it so, the evil planets will not harm him.[32]

I have described these matters for you in the [33]*Bizīdajāt*[33]; ⟨so⟩ I did not mention them [34]for you[34] at the beginning of the book. Take them as [35]guides[35] and work with them, if God, He is exalted, wishes.

If Mars is with the Sun or in the second ⟨place⟩ from it, [36]it (the Sun) subtracts[36] [37]from what I mentioned[37] for it; or ⟨if⟩ it is eastern, in the eleventh ⟨place⟩, then this king is popular, and there is confusion after him.

Know that a king has ruled for forty years and more than this. This happens [38]if[38] it (the Sun) gives its small years in the control of Saturn, Jupiter, and Mars. Understand this: these are the planets which have a share with it (the Sun) in kings ⟨and⟩ the conclusions of their periods ⟨of rule⟩; their small, middle, and big years effect ⟨them⟩.

Know that Venus, [39]Mercury[39], and the Moon do not strengthen the Sun so that it gives more than its small years in their dynasty. Sometimes it gives ⟨only⟩ a third or a

26 in the eighth from the Sun:add. B; in the eighth from the Sun, and Jupiter:add. V. 27-27 signs:V. 28-28 om. B. 29-29 om. V. 30-30 for every year a sign:B. 31-31 He is powerful and great:V. 32-32 om. V. 33-33 al-r.y.r.y.ḥāt:B; al-b.r.r.jāt:V. *Bizīdajāt* is an Arabized form of the Pahlavī of 'Ανθολογίαι, the title of Vettius Valens' work. Another possible reading is *Zā'irjāt* (see *The Thousands of Abū Ma'shar*, pp. 22 and 79); Ibn al-Nadīm's form of the title of Valens' work is the singular of this: *Zā'irj*. Both al-Qaṣrānī and Sahl ibn Bishr use the term zā'irja to designate the horoscopic diagram. 34-34 om. B. 35-35 days:V. 36-36 om. V. 37-37 om. B. 38-38 om. V. 39-39 om. B.

fourth – until the time when an evil planet leaves it; and, as soon as the evil planet leaves it or conjoins with it, the king dies, if God, He is exalted, wishes. And this: that these two ⟨evil⟩ stars are Saturn[40] and Mars[41]. But know that the Sun used to give its middle years in the ancient dynasty, and every planet aspecting it (the Sun) used to increase it by its (the planet's) small years. Therefore the ⟨first⟩ dynasty was long, but the second was not.

Know that Saturn and Jupiter conjoin in every triplicity; learn what I explained to you about this in the beginning of the book – I mean, the *Kitāb al-qirānāt*[42].

The mighty years of the Sun are [43]1461[43]; the big 120; the middle [44]69[44] years and 6 months; and the small 19.

The mighty years of the Moon are 520; the big [45]108[45]; the middle [46]66 and ½[46]; and the small 25.

The mighty years of Saturn are 265; the big [47]57[47]; the middle 43 years and 6 months; and the small 30.

The mighty years of Jupiter are 427[48]; the big 79; the middle 45 ⟨and ½⟩; and the small 12.

The mighty years of Mars are [49]284[49]; the big [50]66[50]; the middle 40 and ½; and the small15.

The mighty years of Venus are [51]1151[51]; the big [52]82[52]; the middle 45; and the small 8.

The mighty years of Mercury are [53]480[53]; the big [54]76[54]; the middle [55]48[55]; and the small 20.

Examples concerning the accession of kings.

The accession of [56]Muḥammad[56], God bless him and grant him salvation.

The ascendant was Aries, Saturn was in Leo, and Jupiter in Pisces; Saturn was in left trine of the Sun, which was in Aries. And so it (the Sun) gave half of its small years; this is 7 years 11 months and 22 days.

The accession of [57]our master[57] Abū Bakr, God is pleased with him.

40 and Jupiter:add. B. 41 and Jupiter:add. V.
42 This is presumably the work summarized by Ibn Hibintā.
43-43 1460:V. 44-44 37:BV. 45-45 180:B. 46-46 37:BV. 47-47 27:B; 48 and a half:add. BV. 49-49 1264:V. 50-50 6:V. 51-51 one hundred and fifty-one:B. 52-52 fifty-two (crossed out), eighty:B; 12:V. 53-53 265:BV. 54-54 eighty-two:B. 55-55 43:BV. 56-56 the prophet of God:B. 57-57 om. B.

The ascendant was Aries, Mars was in Capricorn, Jupiter in Sagittarius, and Saturn in Scorpio. ⟨Because of⟩ the existence of Mars in Capricorn in the tenth (place from the Sun) and of Saturn in the eighth, it (the Sun) gave a ninth of its small years; this is 2 years 3 months and 8 days.

The accession of [58]our master[58] 'Umar ibn al-Khaṭṭāb, God is pleased with him.

The ascendant was Aquarius, and Mars was in it; Saturn was in Sagittarius, Jupiter in Aquarius, and the Sun in Aries; Saturn was in right trine of it (the Sun), and Jupiter and Mars in the eleventh place from it. Therefore it (the Sun) gave three fifths of its small years; this is [59]10[59] years 6 months and [60]17[60] days.

The accession of [61]our master[61] Uthmān ibn 'Affān, God is pleased with him.

. . . Saturn was in Aries and Jupiter in Capricorn . . . ; this is 11 years 11 months and 19 days.

The accession of[62] the Commander of the Faithful, 'Alī ibn Abī Ṭālib, God is pleased with him, and his face is noble with compassion.

Saturn was in Virgo, Jupiter in Capricorn, and Mars in Leo. From the Sun to Mars ⟨are 5 signs; this is⟩ 5 years.

The accession of Mu'āwiya [63]ibn Abī Sufyān[63].

Saturn was in Scorpio and Jupiter in Taurus. From the fact, that Saturn was in the eighth ⟨place⟩ from the Sun, which was in Aries, and that Jupiter was in the second ⟨place from⟩ the Sun – that is, in Taurus – in opposition to Saturn, it (the Sun) gave its small years; this is 19 years 3 months and 22 days.

The accession of Yazīd ibn Mu'āwiya, [64]God curse and dishonor him[64].

The Sun was in Aries and Saturn in Gemini. From the Sun to Saturn are 3 signs; he ruled 3 years.

⟨The accession of Marwān.⟩

The accession of 'Abd al-Malik ibn Marwān.

The ascendant was Sagittarius, the Sun was in Aries, Saturn in Sagittarius, and Jupiter in Capricorn in the tenth ⟨place⟩ from the Sun; Saturn was in the ninth. It (the Sun) gave three fourths of its small years; this is . . . .

58-58 om. B. 59-59 twenty:V. 60-60 seven:V. 61-61 om. B.
62 our master and lord:add. V. 63-63 om. V. 64-64 om. V.

The accession of al-Walīd ibn 'Abd al-Malik [65]ibn Marwān[65]. [66]The Sun was in Aries[66], Saturn in Taurus, and Mars in Cancer 10°; [67]he was killed[67] when [68]the Sun[68] conjoined with Mars, if God wills: 9 years 7 months and 29 days.

The accession of Sulaymān ibn 'Abd al-Malik.

The ascendant was Virgo, the Sun was in Aries, and Saturn in Virgo; and the Sun was in the house of death. It (the Sun) gave ⟨a seventh of its small years⟩; this is 2 years and a half. Similarly [69]know[69]: if Saturn were in [70]Virgo[70], and it were the house of death, it (the Sun) would give this.

The accession of 'Umar ibn 'Abd al-'Azīz, God is pleased with him.

The Sun was in Aries and Saturn in Libra in the second decan; the ascendant was Virgo. If Saturn is in Libra in the second decan in opposition to the Sun, [71]then it is similar[71] . . . from the ascendant; this is 2 years 5 months and 13 days.

[72]The accession of Yazīd ibn 'Abd al-Malik[72].

The Sun was in Aries, Saturn in Scorpio, and Mars in Pisces; ⟨the ascendant was Aries⟩. If one of the two evil planets is in [73]the house of[73] enemies and the other in the house of death, it (the Sun) gives [74]a fourth of[74] its small years; ⟨this is . . . ⟩

The accession of Hishām ibn 'Abd al-Malik.

Saturn was in Sagittarius 25°, Mars in Pisces, and Jupiter in Leo. From the fact [75]that Saturn[75] was in right trine of the Sun, which was in Aries, once it (the Sun) had traveled 20°, it (the Sun) gave its small years; this is 19 years 8 months and 9 days.

The accession of [76]al-Walīd ibn[76] Yazīd.

The Sun was in Aries, Mars in Taurus, Saturn in Leo, and Jupiter in Aries. He ruled as many years as ⟨there are signs⟩ between the Sun and Mars; this is 1 year 2 months and 21 days.

The accession of Yazīd ibn al-Walīd ibn 'Abd al-Malik.

[77]. . . ; 1 month and 9 days.

⟨The accession of Marwān⟩.

65-65 om. V. 66-66 The ascendant was Sagittarius, the Sun was in Aries, Saturn in Sagittarius, and Jupiter in Capricorn in the tenth: V. 67-67 om. B. 68-68 the degree of the Sun: B. 69-69 he says: V. 70-70 the ascendant: BV 71-71 om. V. 72-72 rep. B. 73-73 om. B. 74-74 om. V. 75-75 om. V. 76-76 om. V. 77-77 om. V.

The accession of al-Saffāḥ al-'Abbāsī.

The Sun was in Aries[77], Saturn in Scorpio, and Jupiter in Scorpio. From the fact that Saturn was with Jupiter in the eighth ⟨place from the Sun⟩, it (the Sun) gave a fourth of its small years; this is 4 years 8 months and 2 days.

The accession of Abū Ja'far [78]al-Manṣūr al-'Abbāsī[78].

The Sun was in Aries and Saturn in Capricorn. From the fact that Saturn was in the tenth ⟨place⟩ [79]from the Sun[79], it (the Sun) gave its small years, and Saturn increased them with [80]30[80] months in the measure of its small years; this is . . .

The accession of al-Mahdī [81]al-'Abbāsī[81].

The Sun was in Aries and Saturn in Libra in the first decan in opposition to the Sun. So it (the Sun) gave half of its small years; this is 10 years [82]1 month and 5 days[82].

The accession of Mūsā ibn al-Mahdī [83]al-'Abbāsī[83].

The Sun was in Aries and Saturn in Capricorn 29°. There were 61° between Saturn and [84]the Sun[84]; so it (the Sun) gave 6 months and 15 days.

The accession of Hārūn al-Rashīd.

The Sun was in Aries and Saturn in Aquarius, the eleventh ⟨place⟩ from the Sun. It (the Sun) gave its small years complete, ⟨and Saturn increased them by . . . ⟩; this is 23 years 2 months and 16 days. [85]Fulfillment be his and good fortune[85].

We know of several attempts made in the eighth and ninth centuries to write astrological histories of the Caliphate. One can mention, as earlier than Māshā'allāh, pseudo-Stephanus of Alexandria[1] and the anonymous summarized by al-Sijzī[2]. We also have a long series of horoscopes of the installation of caliphs in the *History* of al-Ya'qūbī and fragments of astrological histories by al-Khwārizmī and al-Battānī, all of which will be discussed elsewhere.

78-78 om. V. 79-79 om. B.
80-80 3:B. 81-81 om. V. 82-82 om. B.
83-83 om. V. 84-84 Capricorn:B. 85-85 So it ends. Glory to God, the lord of the two worlds; and God bless him and us, and Muḥammad and 'Alī; and God go with him and save him safe and sound:V.

1. Ed. H. Usener, *De Stephano Alexandrino* (Bonn, 1880), pp. 17-32, and in his *Kleine Schriften* (Leipzig-Berlin, 1914), vol. 3, pp. 226-289.
2. D. Pingree, *The Thousands of Abū Ma'shar* (Studies of the Warburg Institute, 30; London, 1968), pp. 78-121.

Māshā'allāh's horoscopes are all of the vernal equinoxes of the years in which the several caliphs came to power. The fact that he knows the date of the death of Hārūn al-Rashīd indicates that he wrote this little work shortly after 809. He has saved himself (and us) a great deal of labor by restricting his discussion to the three superior planets and the Sun. The *Zīj al-Shāh* longitudes are computed according to the parameters discovered earlier in this book.

1. The Prophet (7^y 11^m 22^d)

	Text	*Zīj al-Shāh*	Computation (19 March 623)
♄	♌	♌ 4;57	♌ 6
♃	♓	♓ 20;30	♓ 24
♂			
☉	♈ 0;1	♈ 0;1	♈ 1
H	♈		

2. Abū Bakr (2^y 3^m 8^d)

	Text	*Zīj al-Shāh*	Computation (19 March 632)
♄	♏	♏ 28;19	♐ 2
♃	♐	♐ 29;10	♑ 1
♂	♑		♑ 26
☉	♈ 0;1	♈ 0;1	♈ 1
H	♈		

3. 'Umar (10^y 6^m 17^d)

	Text	*Zīj al-Shāh*	Computation (19 March 634)
♄	♐	♐ 19;54	♐ 24
♃	♒	♒ 27;49	♓ 0
♂	♒		♒ 14
☉	♈ 0;1	♈ 0;1	♈ 1
H	♒		

4. 'Uthmān (11^{y} 11^{m} 19^{d})

	Text	*Zīj al-Shāh*	Computation (19 March 644)
♄	♈	♈11;15	♈16
♃	♑	♑ 3;20	♑ 6
♂			
☉	♈0;1	♈ 0;1	♈ 2
H			

5. 'Alī (5^{y})

	Text	*Zīj al-Shāh*	Computation (19 March 656)
♄	♍	♍22;16	♍ 26
♃	♑	♑7;31	♑ 10
♂	♌		♍ 6
☉	♈0;1	♈0;1	♈ 2
H			

6. Mu'āwiya (19^{y} 3^{m} 22^{d})

	Text	*Zīj al-Shāh*	Computation (20 March 661)
♄		♏ 23;0	♏27
♃		♉ 29;28	♊ 3
♂			
☉	♈0;1	♈0;1	♈ 3
H			

7. Yazīd (3^y)

	Text	*Zīj al-Shāh*	Computation (20 March 679)
♄	♊	♊24;21	♊27
♃			
♂			
☉	♈0;1	♈0;1	♈2
H			

8. ʻAbd al-Malik ($–^y$)

	Text	*Zīj al-Shāh*	Computation (20 March 692)
♄	♐	♐9;43	♐15
♃	♑	♑19;48	♑23
♂			
☉	♈0;1	♈0;1	♈3
H	♐		

9. Al-Walīd (9^y 7^m 29^d)

	Text	*Zīj al-Shāh*	Computation (20 March 705)
♄	♉	♉7;32	♉13
♃			
♂	♋10°		♎20(!)
☉	♈0;1	♈0;1	♈3
H			

10. Sulaymān (2½^{y})

	Text	*Zīj al-Shāh*	Computation (20 March 714)
♄	♍	♍ 9;42	♍15
♃			
♂			
☉	♈ 0;1	♈ 0;1	♈ 3
H	♍		

11. 'Umar (2^{y} 5^{m} 13^{d})

	Text	*Zīj al-Shāh*	Computation (20 March 717)
♄	♎10-20	♎18;15	♎23
♃			
♂			
☉	♈ 0;1	♈ 0;1	♈ 3
H	♍		

12. Yazīd (–^{y})

	Text	*Zīj al Shāh*	Computation (20 March 719)
♄	♏	♏12;10	♏17
♃			
♂	♓		♈5
☉	♈ 0;1	♈ 0;1	♈2
H			

13. Hishām (19^y 8^m 9^d)

	Text	*Zīj al-Shāh*	Computation (20 March 723)
♄	♐ 25	♐ 25;54	♑ 2
♃	♌	♌ 19;31	♌ 25
♂	♓		♉ 5
☉	♈ 0;1	♈ 0;1	♈ 3
H			

14. Al-Walīd (1^y 2^m 21^d)

	Text	*Zīj al-Shāh*	Computation (20 March 743)
♄	♌	♌ 19;50	♌ 25
♃	♈	♈ 3;9	♈ 6
♂	♉		♉ 28
☉	♈ 0;1	♈ 0;1	♈ 3
H			

15. Al-Ṣaffāḥ (4^y 8^m 2^d)

	Text	*Zīj al-Shāh*	Computation (20 March 749)
♄	♏	♏ 18;7	♏ 24
♃	♏	♏ 7;26	♏ 11
♂			
☉	♈ 0;1	♈ 0;1	♈ 3
H			

16. Al-Manṣūr (—^{y})

	Text	*Zīj al-Shāh*	Computation (20 March 754)
♄	♑	♑ 11;46	♑ 19
♃			
♂			
☉	♈ 0;1	♈ 0;1	♈ 3
H			

17. Al-Mahdī (10^{y} 1^{m} 5^{d})

	Text	*Zīj al-Shāh*	Computation (20 March 775)
♄	♎ 0-10	♎ 6;14	♎ 13
♃			
♂			
☉	♈ 0;1	♈ 0;1	♈ 3
H			

18. Mūsā (6^{m} 15^{d})

	Text	*Zīj al-Shāh*	Computation (20 March 785)
♄	♑ 29	♑ 27;46	♒ 7
♃			
♂			
☉	♈ 0;1	♈ 0;1	♈ 4
H			

19. Hārūn al-Rashīd ($23^y\ 2^m\ 16^d$)

	Text	*Zīj al-Shāh*	Computation (20 March 786)
♄	♒	♒8;30	♒18
♃			
♂			
☉	♈0;1	♈0;1	♈3
H			

From these figures it is clear that Māshā'allāh has used the *Zīj al-Shāh* of Anūshirwān in computing these horoscopes. The dates he gives are probably of little historical importance; but it may be interesting to compare them with Bīrūnī's and those of the work summarized by al-Sijzī:[3]

Caliph	Bīrūnī	al-Sijzī	Māshā'allāh
1. Muḥammad		622	623 ($7^y\ 11^m\ 22^d$
2. Abū Bakr		632	632 ($2^y\ 3^m\ 8^d$)
3. 'Umar I		634	634 ($10^y\ 6^m\ 17^d$)
4. 'Uthmān		644	644 ($11^y\ 11^m\ 19^d$
5. 'Alī		656	656 (5^y)
6. Mu'āwiya I	($19^y\ 3^m\ 25^d$)	661	661 ($19^y\ 3^m\ 22^d$)
7. Yazīd I	($3^y\ 8^m$)	680!	679 (3^y)
8. Mu'āwiya II	($3^m\ 22^d$)	683	
9. Marwān I	($8^y\ 9^m$)	684!	
10. 'Abd al-Malik	($13^y\ 6^m\ 8^d$)	692	692
11. Al-Walīd I	($9^y\ 7^m\ 29^d$)	705	705 ($9^y\ 7^m\ 29^d$)
12. Sulaymān	($2^y\ 7^m\ 29^d$)	714	714 ($2\frac{1}{2}^y$)
13. 'Umar II	($2^y\ 5^m\ 13^d$)	717	717 ($2^y\ 5^m\ 13^d$)
14. Yazīd II	($4^y\ 0^m\ 1^d$)	719	719
15. Hishām	($19^y\ 8^m\ 9^d$)	723	723 ($19^y\ 8^m\ 9^d$)

3. For Bīrūnī's chronology of the caliphs, see *Chronology*, ed. K. Garbers, in *Documenta Islamica Inedita*, ed. J. W. Fück (Akademie Verlag, Berlin, 1952), pp. 60-66.

16. Al-Walīd II	($1^y 2^m 21^d$)	742	742	($1^y 2^m 21^d$)
17. confusion	($2^m 25^d$)	744!		
18. Yazīd III	($2^m 9^d$)			
19. Ibrāhīm	($2^m 11^d$)			
20. Marwān II	($5^y 2^m 0^d$)			
21. Al-Ṣaffāḥ	($4^y 8^m 2^d$)	749	749	($4^y 8^m 2^d$)
22. Al-Manṣūr	($21^y 11^m 22^d$)	754	754	
23. Al-Mahdī	($10^y 1^m 17^d$)	775	775	($10^y 1^m 5^d$)
24. Al-Hādī	($1^y 1^m 23^d$)	785	785	($6^m 15^d$)
25. Al-Rashīd	($23^y 2^m 16^d$)	786	786	($23^y 2^m 16^d$)

Bīrūnī's and Māshā'allāh's agreement concerning the lengths of the reigns of 11, 13, 15, 16, 21, and 25 indicates that, at least in part, they shared a common source. Perhaps this source was truth.

APPENDIX 3

Māshā'allāh's *Kitāb al-mawālīd*

Since the Arabic original of this work appears to be lost, save for what is quoted by Māshā'allāh's pupil Abū 'Alī al-Khayyāṭ (the only manuscript of the Arabic original of his work seems to be Hamidiye 856), I edit here the Latin translation based on the apparently unique manuscript: Paris BN Latin 7324 ff. 73-76. Words enclosed within pointed brackets are my additions. D.P.

⟦f 73a⟧ Incipit Liber Mesellae De nativitatibus.

Et primo an sit puer abla⟨cta⟩tus aut non.

Dixit Messeallah quod inter omnes libros astronomie non invenitur utilior libro nativitatum neque tam bonus in iudiciis. Qui fuit prudens in ea inveniet prudenciam et sapienciam in eo et delectabit in eius pericia. Inprimis sciri oportet si ablactatus fuerit puer aut non; et hoc scies per ascendentem, scilicet per dominos triplicitatis, et per dominos domus Solis, si nativitas diurna fuerit, ⟨aut domus Lune, si fuerit nocturna⟩; et a domino coniunctionis aut domino impletionis si fuerit natus post ⟨coniunctionem aut⟩ impletionem. Et si fuerit ex coniunctione ⟨et impletione⟩, aspicies a dominis triplicitatis; insuper hoc aspicies Iovem et Venerem. Quod si nativitas diurna fuerit, aspicies ex stellis diurnis; et si nocturna, ex stellis nocturnis. Et incipies aspicere ad dominum triplicitatis prime domus, id est ascendentis, et ad dominum secundum et ad dominum tercium; hii sunt domini prime triplicitatis. Qui

2. ablatus 4. libro *written above line* 7. ablatus, *corr. to* ablactatus 8. triplicitatum 10. et . . . domino[2] *written between lines* [aut domino] et dominus et 12. triplicitatum 15. a domino triplicitatum

si fuerint liberi a malis in ascendente, scilicet aut in 10ª domo seu in undecima vel in 5ª, vivet natus. Qui si fuerint cadentes ab angulis et fuerint infortune, aspicies ad dominos triplicitatis domus in quo Sol tunc fuerit si fuerit nativitas diurna; si vero nativitas nocturna fuerit, aspicies ad dominos triplicitatis domus in quo Luna tunc fuerit. Qui si in bono loco et salvi a malis fuerint, vivet natus. Qui si in malis locis fuerint et infortune, aspicies ad dominos triplicitatis domus in qua Pars Fortune tunc fuerit. Qui si fuerint in bonis locis et liberi a malis, vivet natus. Quod si nativitas diurna fuerit, aspicies ad Partem Fortune; et si nocturna fuerit, ad Lunam aspicies. Quae si fuerit in malis locis, aspicies ad dominos triplicitatis domus coniunctionis Solis et Lune, si nativitas circa coniunctionem fuerit; vel aspicies ad dominos triplicitatis domus inplecionis, si nativitas circa inplecionem fuerit. Qui si in angulis fuerint et a malis liberi, vivet natus. Qui si mali et inpediti fuerint, aspice ad Iovem qui est particeps in nativitatibus. Qui si fuerit in angulo aut in succedente anguli et liber a malis, vivet; si vero in malo loco fuerit et impeditus a malo, aspicies Venerem. Quae si fuerit in angulo aut in succedente anguli et a malis libera, vivet; si vero in malo loco et ab infortunatis inpedita fuerit, aspicies Lunam. Quae si in ascendente fuerit aut in 10ª domo et libera a malis fuerit aut iuncta diurne stelle si nativitas diurna fuerit aut iuncta stelle nocturne si nativitas nocturna fuerit et a malis libera, vivet; que si mala fuerit, aspicies ad almutez. Et scies per dominos triplicitatis ascendentis; et si nativitas diurna fuerit, per dominos triplicitatis Solis; et si nativitas nocturna fuerit, per dominos triplicitatis Lune; et similiter per dominos triplicitatis Partis Fortune; et per dominos triplicitatis ⟨domus⟩ coniunctionis Solis et Lune si nativitas circa coniunctionem fuerit; vel per dominos triplicitatis domus impletionis si nativitas ante coniunctionem fuerit. Qui si in angulis fuerint aut in succedentibus angulorum et a malis liberi, vivet; qui si in malis ⟨locis⟩ et impediti fuerint,

2. fuerint[1]] fuerit | cadens 3. a dominus triplicitatum 5. a domino triplicitatum 7. in infortune 8. triplicitatum 12. triplicitatum 14. triplicitatum 17. quod | partices 18. angulis | aut [succedente *canceled*] in succedentibus | angulorum 19. in malis locis 20. angulis | succedentibus angulorum 21. in malis locis 22. fuit 26. triplicitatum 27. triplicitatum 28. triplicitatum 29. triplicitatum 30. triplicitatum 32. triplicitatum | impleticionis 33. fuerit

morietur. Deinde aspicies planetam qui fuerit almutez et cui dispositionem suam communicaverit et qui gradus interfuerint. Qui si fuerit in signo fixo, dabis unicuique gradui annum; et si fuerit in signo communi, dabis unicuique gradui mensem; et si fuerit in signo mobili, dabis unicuique gradui diem. Si vero almutez in cadente fuerit et malus planeta in ascendente et Luna iuncta malo, vivet tot tempus quot gradus fuerint – id est, si ipse qui receptor dispositionis inpeditus fuerit et erit in signo fixo, erunt anni; et si in signo communi, erunt menses; et si in signo mobili, erunt dies. Deinde aspice ad gradus Solis vel Lune. Ex quarto aspectu si se aspexerint, cum almuptez aliquis eorum, aut habuerint coniunctionem in una domo vel aspexerint se ex oppositione, erit malum nisi fuerint ibi gradus atascir. Et si ibi fuerint aliquis graduum atascir, vivet tot ⟦f 73b⟧ annis aut mensibus seu diebus, ut supradictum est, quot gradus fuerint. Si vero almutaz iunctus fuerit malo planete ex 4° aspectu vel ex 7° et aliquis fortuna non aspexerit eum, parum vivet. Quod si Luna fuerit coniuncta inter duos malos et unus fuerit in ascendente, alius in 7°, et Luna in angulo mala, morietur. Set domini triplicitatis ascendentis vel domini triplicitatis domus in quo fuerit Sol et domini triplicitatis domus in qua fuerit Pars Fortune et domini triplicitatis domus coniunctionis vel implecionis impediti fuerint vel in cadentibus et aliquis planeta fuerit in fortitudine, habebit graves egritudines; et si domini triplicitatis domus ascendentis fuerint in cadentibus, morietur; et deterius erit si unus predictorum fuerit Saturnus in nativitate nocturna vel fuerit Mars in nativitate diurna, scilicet in uno angulorum. Quod si Luna recepta fuerit et dominus ascendentis in bono loco, vivet eritque honoratus et multos habebit fratres; quod si receptio ibi non fuerit, designat inopiam. Si Pars Fortune fuerit cum Luna et aspexerit Venerem in nativitate nocturna aut Iovem in nativitate diurna, designat altitudinem et vitam sicut si Pars Fortune in bono loco fuerit. Omnis planeta qui indicat super nativitatem et fuerit orientalis in nativitate

1. almutep 2. *after* interfuerint *MS adds* glossa a receptione [*read* receptore] dispositionis 6. almutep | cadentibus 7. fuerit | *after* fuerit *MS adds and cancels* id est receptus si impeditus 13. domorum 15. menses | dies 16. almutap 17. fortune 18. aspexerint 20. moritur | triplicitatum 21. triplicitatum 22. triplicitatum 23. triplicitatum 25. triplicitatum 26. fuerit 33. jupiter

diurna et in signo masculino aut occidentalis in nativitate nocturna et in signo feminino, erit fortitudo planete bona et testimonium eius bonum, et designat sublimitatem nati. Quod si dominus ascendentis aut Luna in malo fuerit et dominus domus Lune in angulo, designat mortem. Cumque sciveris natum non diu vivere, facies atasir a gradibus ascendentis usque ad planetam malum qui impedit, et dabis unicuique signo mensem; set si natus ab hiis mensibus evaserit, tot annos vivet quot menses fuerint predicti. Deinde aspice ad dominum 5 domus; qui si in bono loco fuerit, predic⟨tum⟩ periu⟨dican⟩dum bon⟨um⟩; si in ⟨malo⟩, angustiabit animum suum in paupertate.

Quando fuerit yles in nativitate vel non.

Capitulum de yle in scientia vite, cum nativitas pueri designat vitam.

Et cum hoc scire volueris, dirige ylem in nativitate diurna a Sole qui si fuerit in angulo aut in succedente anguli et in signo masculino aut in 4^a parte masculina – id est a 7^a domo in antea – et aspexerit eum dominus domus sue aut dominus termini eius aut dominus sue exaltationis aut dominus sue triplicitatis aut sue faciei, erit yles; et si unus horum non aspexerit eum, non erit yles. Deinde Lunam aspicies que si fuerit in angulo aut in succedente anguli et in signo feminino aut in 4^a feminina et aspexerit eam sicut dixi in Sole, accipies eam pro yle. Quod si Sol aut Luna non fuerit yles, aspicies ad dominum domus coniunctionis vel inpletionis; si non fuerit yles, aspicies ad dominum domus Partis Fortune; et si hunc ylem non inveneris, pones gradum ascendentis pro yle, si dominus ascendentis aspexerit ascendens. Set si omnes predicti defecerint, non erit yles.

De noticia temporis ex alcoden.

Capitulum de alcoden per quod scitur computatio vite.

2. femenino 5. in angulo *written between lines* 6. *above* atasir *is written* id est directum 10-11. predictum . . . malo *written in margin by a second hand; it is partly missing from the photograph of the MS* 15. hoc *written above line* | yles 16. angulis | succedentibus angulorum 18. sue *written in margin* 19. *after* dominus *MS adds* domus 22. angulis | succedentibus angulorum 24. fuerint 25-26. si fuerint yles et non fuerint yles 26. *before* aspicies *MS adds and cancels* ad 27. gradus 31. de yle *canceled* ex alcoden

Cumque inventum fuerit yles, aspicies alcoden; et aspicies ylem et ad dominum termini eius et ad dominum sue triplicitatis et ad dominum domus sue et ad dominum sue exaltationis et ad dominum sue faciei; et horum qui aspexerit ylem erit alcoden. Si vero unus planetarum aut duo vel tres aspexerint ad ylem, ipse planeta qui fuerit plus in sua fortitudine et propinquior gradibus erit alcoden. Scias quia cum Sol fuerit in Ariete vel Leone, erit alcoden, erit yles; similiter si Luna fuerit in Tauro aut in Cancro, ipsa erit yles et alcoden, si aspexerint eam aut non; similiter in Sole. Et cum alcoden inveneris, aspice ipsum; si fuerit in angulo ex gradibus aut in domo sua aut in exaltatione seu triplicitate salvus ab inpedimento scilicet de retrogradatione aut de adustione Solis, dabis ei annos maiores ipsius planete; qui si fuerit in succedente anguli et salvus a malis, dabis ei annos mediocres; et si fuerit in cadente anguli et ibi aliquam dignitatem non habuerit, dabis ei annos minores. Et scias quod augmentum annorum planete vel diminucionem eorum non est nisi ex fortitudine planete seu de ⟦f 74a⟧ bilitate eius. Set si ipse planeta fuerit orientalis et in bono esse, dabis maiores; et si orientalis non fuerit et habuerit malum aspectum cum aliqua, dabis ei annos minores; et si fuerit occidentalis et malum habuerit aspectum cum aliqua et fuerit retrogradus, dabis ei ebdomadas quot fuerint anni minores; qui si in malo fuerit loco in quo non possit esse deterior cum retrogradus et de levioribus planetis, dabis ei dies tot quot anni fuerint minores ipsius planete. Et scias quod Capud Draconis cum fuerit in uno signo cum planeta qui fuerit almutez infra 12 gradus vel plus ante vel retro addit quartam partem ipsorum annorum illius planete qui fuerit almutez; et si fuerit propinquius graduum, tamen erit utilior. Quod si Cauda Draconis ibi fuerit, minuit quartam partem annorum; et si cum Sole aut Luna fuerit, gradualiter nihil minuit; si vero Sol fuerit alcoden et elongatus fuerit ab ea, minuet de annis. Dixit Ptolomeus: Capud cum fortunis addit fortunam et Cauda minuit de annis. Quod si cum Sole et Luna fuerit Capud vel Cauda, apparebit fortitudo eorum

4. aspexerint 6. yles 8. est *written above* erit[1] | *after* yles *MS adds* alcoden 14. succedentibus angulorum 15. eis 16. cadentibus angulorum | habuerint 17. eis | aucmentum 23. annos 28. almutep 30. almutep 35. sol

sive in bono sive in malo; et in Luna forcius. Set si alcoden in malo loco fuerit, minuet de annis. Quod si Iupiter cum Venere in ascendente fuerit, unusquisque horum addit in nativitate annos minores nisi infortune eos inpedierint et Luna similiter in malo esse. Si vero predicte fortune scilicet Iupiter et Venus unus horum fuerit dominus domus mortis et in ascendente fuerit, morietur ante parum vivet.

Quot annos addunt planete alcoden.

Capitulum ad sciendum quid addunt planete vel minuunt.

Et cum sciveris quot annos debes ponere per alcoden et scire volueris quot minuuntur vel adduntur, aspicies alcoden; si fuerit cum eo fortuna et aspexerit cum eo ex sextili aspectu seu aliquo bono et si in bono loco fuerit, augebit annos minores ipsius fortune; si vero fortune qui alcoden aspexerint fuerint debeles, pro annis minoribus dabis ei tot menses; et ⟨si⟩ fuerit retrogradus fortuna planeta qui alcoden aspexerit et malus eam inpedierit, dabit ei tot ebdomadas quot anni minores ipsius planete retrogradi; si vero alcoden cum domino domus mortis fuerit et aliquis stella mala eum inpedierit, erunt dies quot anni minores fuerint. Quod si stella infortuna cum alcoden fuerit et stella illa erit receptor gradus alcoden et aspexerint se ex quarto aspectu aut ex oppositione et habuerint coniunctionem cum domino domus mortis, erunt hore tot quot fuerint anni minores ipsius alcoden. Si vero Mercurius in bono loco fuerit et aspexerit alcoden bono aspectu, augentur ei anni minores; et si e converso fuerit, minuentur ei tot anni quot fuerint minores. Et forcior planetarum aspectuum est aspectus Martis cum alcoden. Sed si scire volueris certitudinem mortis, aspicies ad malum qui inpedierit alcoden; et cum alcoden pervenerit ad illos gradus, morietur.

Cuius voluntatis est natus.

Capitulum ad sciendum voluntatem nati.

Aspice ad dominum ascendentis et Mercurium, qui indicat super locutionem nati; qui si fuerit fortis et in signo mobili, indicat eum bonam locutionem habere et honoratum et timentem Deum; et si fuerit in signo communi, indicat eum habere paucam sapienciam et iracibilem, et fere non credit consilium alterius; et si in signo fixo fuerit, indicat eum fore

3. *after* fuerit *MS adds* et 4. impedierit 18. retrogradus 31. volutatis 34. nati *written between lines*

honoratum et per veritatem et bonitatem et consilium in vita sua, et in omni modo consilium eius erit verissimum et liberabit consilia inpedita ab inpedimentis. Qui si fuerint orientales et in angulis aut in succedentibus angulorum, indicat bonum esse et acutum ingenii, et quodcumque facere voluerit sine impedimento faciet; qui si fuerint occidentales et ab angulis cadentes, indicat eum malivolum corde, et societatem vilium personarum, et erit nimis iracibilis. Et omne quod predictum est ex domino ascendentis et Luna; qui si in bono loco fuerint, dic bonum; si autem e converso, unaquaeque stella dominium optinet in corpore humano. Si vero dominus ascendentis fuerit Sol et in bono loco salvus a malis, indicat altitudinem et honorem et dominium; qui si malus et in cadenti loco fuerit, indicat eum malum habere spiritum et paucam fortitudinem et parum acquisitionis lucri. Sed si Luna fuerit domina ascendentis et ⟨in⟩ bono loco libera a malis, indicat eum honestum et verecundum; et si in malo loco fuerit, designat eum ⟦f 74b⟧ esse inverecun⟨dum⟩ et tristem, et ex ore suo procedet . . . . per quem corpus verberabitur et percutietur. Saturnus si fuerit dominus ascendentis fueritque in bono loco liber a malis, designat eum esse honoratum et graciosum et intentum in orationem et in habilitatem, in consilio prudentem; et si in malo loco fuerit, designat eum esse tristem et lamentantem et deceptorem. Iupiter si fuerit dominus ascendentis et in bono loco et liber a malis, designat sublimitatem et honorem et bonum spiritum; et si in malo loco fuerit, designat eum esse inverecundum et parcum. Mars si dominus ascendentis fuerit et liber a malis, designat eum esse luxuriosum, et erit forcis corde; et si in malo loco fuerit, designat eum esse alhagem, et erit minutor sanguinis aut carnifex. Venus si domina ascendentis fuerit et in bono loco, designat eum esse pulcrum et humilem; et si in malo loco fuerit, designat eum esse calidum; et si mulier fuerit, designat eam esse meretricem. Mercurius si dominus ascendentis fuerit et salvus a malis, designat eum esse sapientem et medicum; qui si dominus ascendentis fuerit et iunctus Saturno, designat eum esse

2. mode | eius *written between lines* 7. cadentibus 12. saluvus 23. prudentum 24. lametantem 28. si in ascendenti 31-32. si in ascendente 34. callidus *after which MS adds and cancels* si filia fuerit 35. si in ascendente 36-37. si in ascendente

balbucium lingua et bonum medicum in forcibus rebus et luxuriosum sive sodomitam; qui si cum Iove iunctus fuerit, designat eum esse bonum et sapientem, et optinet dominium prepositum; sed si iunctus fuerit Marti aut in suo aspectu, designat eum esse regem aut scriptorem regis; qui si iunctus fuerit Veneri, designat eum diligere sapienciam et iudicia, et erit perfidiosus in lege, perpessus in fortitudine aut eius debilitate aut in superbia animi aut in detrimento sue scientie. Scies ex finibus (?) signorum de eorum fortitudine aut debilitate, sicuti docui in libro ⟨De⟩ ix partibus Messalla uti scribitur in Zoelbembris,* Alichel, et Alicberz, etc. Deinde pro filiis regum aspicies ad gradum ascendentis; si fuerit aliqua stella calida ad predictum gradum iuncta aut in medio celi, et fuerit gradus ascendentis lucidus, aut fuerit ibi Sol in nativitate diurna aut Luna in nativitate nocturna, et fuerit natus ex progenie regum, designat esse eum sub potestate regis, et erit sublimis. Quod si nativitas diurna fuerit et Sol in Ariete aut Luna fuerit in sua exaltatione in nativitate nocturna ac in medio celi aut in gradu ascendentis, et erit ascendens de signis qui reges designant, et dominus ascendentis in bono loco, indicat pro regno et pro sublimitate. Sed si duo luminaria iuncta fuerint domino ascendentis in sua exaltatione, designat eum habere forte regnum. Quod si dominus ascendentis iunctus fuerit cum domino 10^{e} domus et fuerint orientales et in sua exaltatione, erit rex potentissimus. Si vero plures stellarum iuncte fuerint cum Iove in medio celi aut in sua exaltatione, indicat eum esse imperatorem; et unaquaque stella quae fuerit almutaz et fuerit sicut predictum est, erit ei super sublimitatem. Quod si dominus triplicitatis ascendentis iunctus fuerit cum domino ascendentis in medio celi et fuerit orientalis, designat eum esse regem. Operare cum stellis diurnis cum Sole, et cum stellis nocturnis cum Luna; et si Sol fuerit in domo sua aut in angulo seu gaudio et stelle se adinvicem aspexerint, designat eum esse regem.

De fortunio et infortunio nati.

1. ligna 3. sapiente 5. scripor 9. fnbus? *or* fulris? 10. *after* partibus *MS adds* de 13. in] ex 14. lucidus *written above* iunctus *which is canceled* 16. proienie 20. designat 22. in] ex 23. fortem 26. *after* plures *MS adds and cancels* planetarum 28. qui | almutap 29. triplicitatum 33. *after* Luna *MS adds* et sicut orientales

*Sahl ibn Bishr.

Capitulum ad sciendum fortunium et infortunium nati in sua nativitate.

Aspices ad ⟨dominos⟩ triplicitatis luminaris cui congruum fuerit: in nativitate diurna Solis et in nocturna Lunae; qui si in angulis fuerint et liberi a malis, indicat fortunam bonam nati in omnibus diebus vite sue. Et si dominus triplicitatis primus fuerit in ascendente a primo gradu usque ad 15, ascendet ad magnas divicias; et cum prior (?) gradibus in angulo fuerit, erit ei utilior. Quod si fuerit in 2° signo a primo gradu usque ad 15, . . . Sed si dominus primus triplicitatis fuerit in bono loco, erit ei bonum in primo tempore; et si dominus secundus triplicitatis, erit in secundo tempore; et si dominus 3us, erit in 3° tempore; et si e converso, erit e converso. Si vero dominus triplicitatis luminaris fuerit cadens et in malo loco, designat eum habere inopiam, scilicet in quo tempore fuerit in prima parte aut in secunda seu in tercia. Set si domini triplicitatis ⟦f 75a⟧ luminaris fuerint in malis locis, et fuerint in angulis fortune, et aspexerit eos ascendens, et non fuerint inpediti, designat eum habere fortunam. Et si luminaria non fuerint inpedita, erit ei utilior. Quod si dominus ascendentis et Luna fuerint in angulis et liberi a malis, indicant eum habere divicias, et magis si recepti fuerint; si vero dominus ascendentis iungetur alicui luminarium in domo sua aut in exaltatione, aut luminaria se iunxerint domino ascendentis, designat eum excelsum in divitiis. Et si Pars Fortune et dominus eius in angulo, scilicet in oriente, et aspexerint ascendens (!), indicat eum esse ditissimum; qui si in cadenti⟨bus⟩ et in malis locis fuerint, designat eum habere detrimentum, et magis nisi aspexerint ascendens domini triplicitatis ascendentis. Quod si fuerint in cadentibus angulorum et iuncti fuerint cum fortunis in angulis, designat eum habere bonum prae detrimentum; cum dominus ascendentis fuerit cadens et in suo detrimento et iunctus fuerit cum planeta qui fuerit in sua exaltatione aut in sua domo, designat eum habere bonum post detrimentum.

3. triplicitatem 4. a sole | ad luna | quae 5. fuerit | libera 7. primus] prime 8. proprior 10. *after* 15 *MS repeats* erit ei utilior 10-11. prime triplicitatum 12. secunde triplicitatum 13. 3us | 3° 14. triplicitatum 17. triplicitatum 17-18. malo loco 20. *after* inpedita *MS adds and cancels* designat 23. alacri | *after* luminarium *MS adds* et similiter 23-24. domibus suis 24. exaltationibus 25. excelsus 27. ditissimus 30. triplicitatum

Appendix 3

Exemplum Messallah

figura nocturna

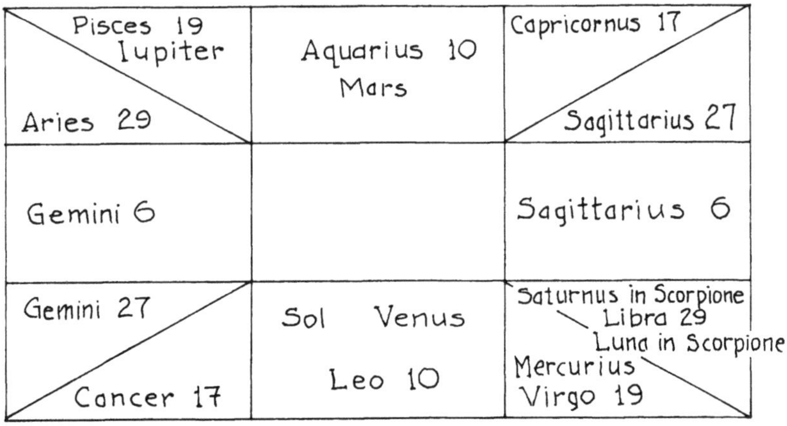

[Horoscope 3.1] Aspexi fortunam huius nati ad dominos triplicitatis Lune quia nativitas fuit nocturna. Et dominus primus triplicitatis erat Mars, Venus 2us, et dominus 3us triplicitatis Luna; et erant cadentes. Iudicavi eum paupertati, et ad tempus vidi eum mendicum; et obiit in miseria et in paupertate. Et hoc totum patet in ista figura.

15. triplicitatum 16. prime | ♀ 2us *written in margin* | dominus 3°

figura diurna

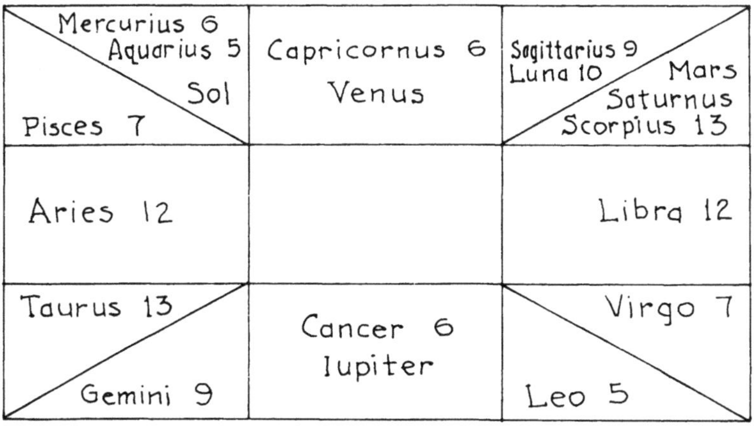

[Horoscope 3.2.] Aspexi fortunam istius nati a dominis triplicitatis Solis, ideo quia nativitas fuit diurna; fuitque dominus triplicitatis Saturnus, et secundus Mercurius, et ipse fere in angulo; et unus in descendente et alius in medio celi. Deinde iudicavi eum habere bonitatem et pacienciam multam; et ita fuit bonus semper.

13. triplicitatum 15. angulis | ascendente

figura diurna

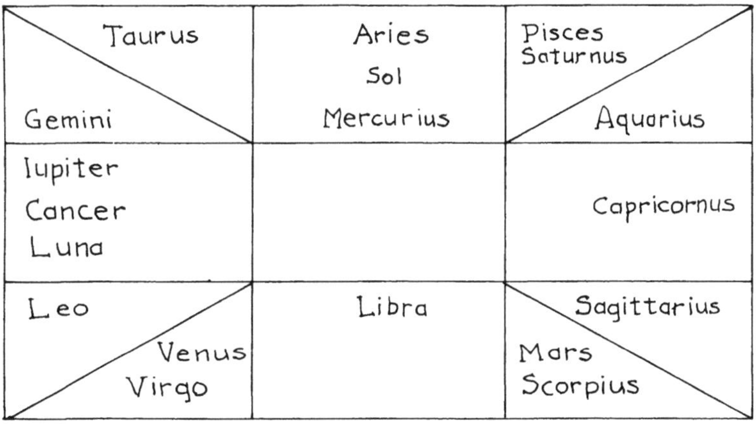

[Horoscope 3.3] Aspexi ad fortunam nati ex dominis triplicitatis Solis; et dominus primus triplicitatis est Sol, et secundus Iupiter, et dominus 3us est Saturnus. Et hos inveni in angulis et in suis exaltationibus; designabat eum habere fortunam et sublimitatem, et ideo non nocuit ei Saturnus quia erat in domo Iovis et aspiciebat se bono aspectu.

14. *after* 3us *MS adds and cancels* triplicitatis

Kitāb al-mawālīd

⟦f. 75b⟧ nativitas nocturna

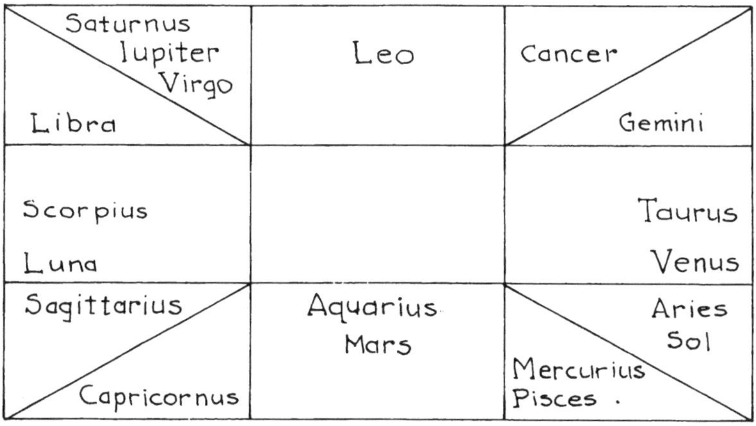

[f. 75b. Horoscope 3.4.] Aspexi fortunam eius ad dominos triplicitatis Lune; et dominus primus triplicitatis est Mars, dominus secundus est Venus, dominus 3us Luna. Et hos inveni in angulis; ideo iudicavi eum quia ascenderet in sublimitatem nimis donec adquireret coronam auream; et fuit ita.

13. triplicitatum

Appendix 3

nativitas nocturna

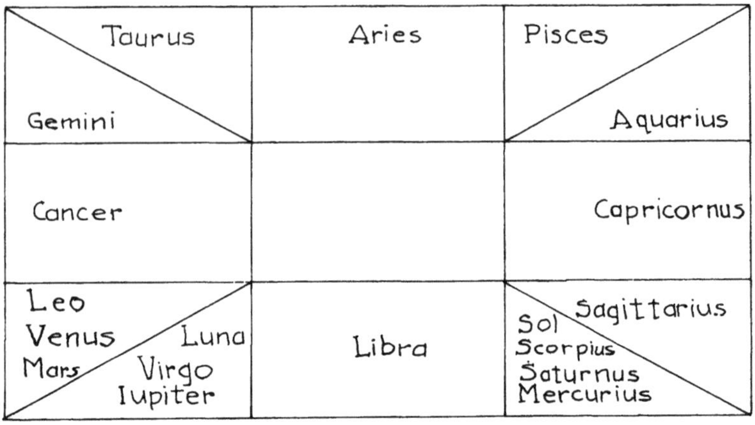

[Horoscope 3.5.] Aspexi ad fortunam istius nati a dominis triplicitatis Lune, qui sunt Mercurius primus, Saturnus secundus, Iupiter tertius. Et hii erant cadentes. Iudicavi super eum miseriam et inopiam, et ita fuit.

13. iupiter tertius *written between lines*

Kitāb al-mawālīd

nativitas diurna

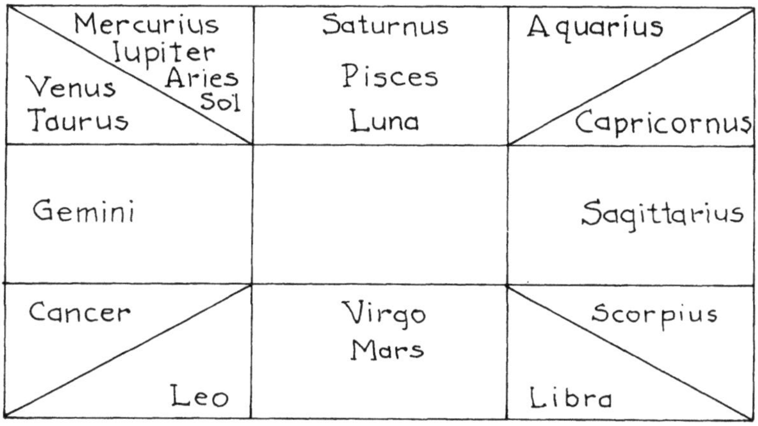

[Horoscope 3.6.] Aspexi ad fortunam istius nati ex dominis triplicitatis Solis, quorum primus est Sol, dominus secundus Iupiter, tercius Saturnus. Et hos inveni in angulis. Iudicavi eum habere sublimitatem et celsitudinem et honorem in omnibus vite sue diebus; et ita fuit.

8. geminis

Appendix 3

nativitas nocturna

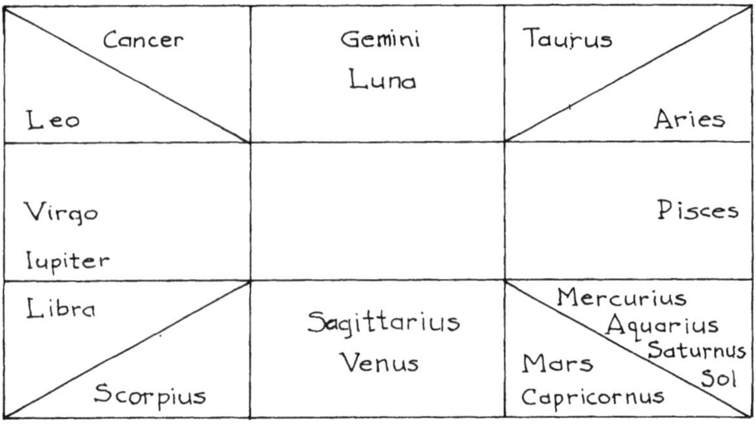

[Horoscope 3.7.] Aspexi ad dominos triplicitatis Lune, quorum primus Mercurius, dominus secundus Saturnus, dominus tercius triplicitatis Iupiter. Et hos inveni cadentes ab angulis et in malis locis. Iudicavi eum esse miserum et mendicum.

14. tercie

Kitāb al-mawālīd

nativitas diurna

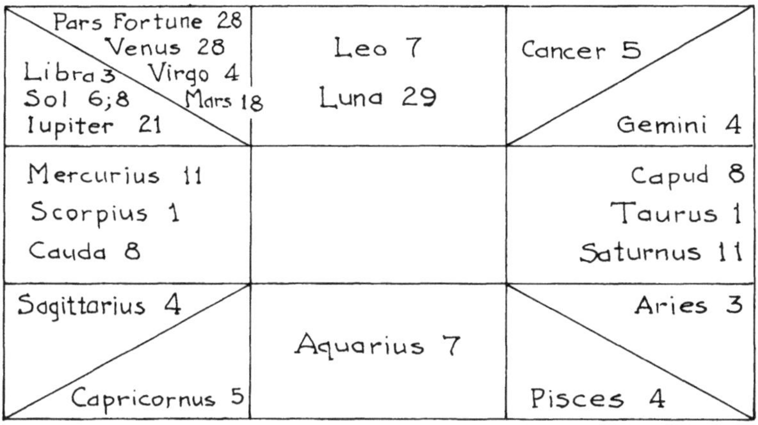

[Horoscope 3.8.] Aspexi ad fortunam istius nati ex dominis triplicitatis Solis, quorum primus erat Saturnus, secundus Mercurius, 3us Iupiter. Inveni hos aspicien⟨tes⟩ ex oppositione; et ⟨Mercurius⟩ iunctus cum Cauda, et Saturnus cum Capite, et Iupiter cadens, et Pars Fortune cum Marte, et dominus Partis Fortune iunctus cum Cauda, et fuit in inpletione Saturnus. Deinde iudicavimus perstultum et miserum; et istud totum patet in ista figura.

13. ex do dominis 16. iuncti

Appendix 3

nativitas nocturna

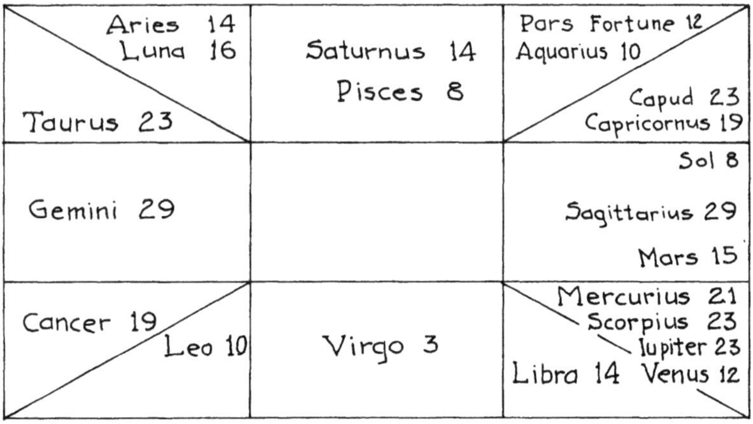

[Horoscope 3.9.] Aspexi ad fortunam istius nati ex dominis triplicitatis Lune, quorum primus est Sol, secundus Iupiter, tercius Saturnus. Et hos inveni cadentes in 6ª domo. Indicat eum ideo pro malo et parum bonum pro testimonio Partis Fortune, quae indicat pro malo.

15. triplicitatum

[f. 76a] nativitas nocturna

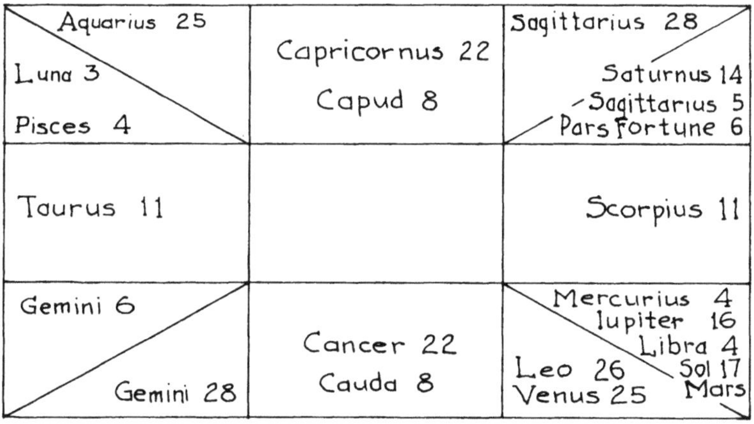

[f. 76a. Horoscope 3.10.] Aspexi ad fortunam istius a dominis triplicitatis Lune, quorum primus est Mars, et erat sub radiis Solis in 4° aspectu. Indicat pro amissione et malo in primo tempore vite sue, et in 2ª parte vite sue indicat pro bono ideo quia Venus erat dominus secundus, eratque in bono esse; et designat quod in secundo tempore vite sue habebit bonum prae multum laborem. Tamen Iupiter et Mercurius erant in VIª domo ex Parte Fortune; designat eum habere honorem et prospectum. Ita fuit.

14. domino 17. *after* secundus *MS adds and cancels* triplicitatis

nativitas nocturna

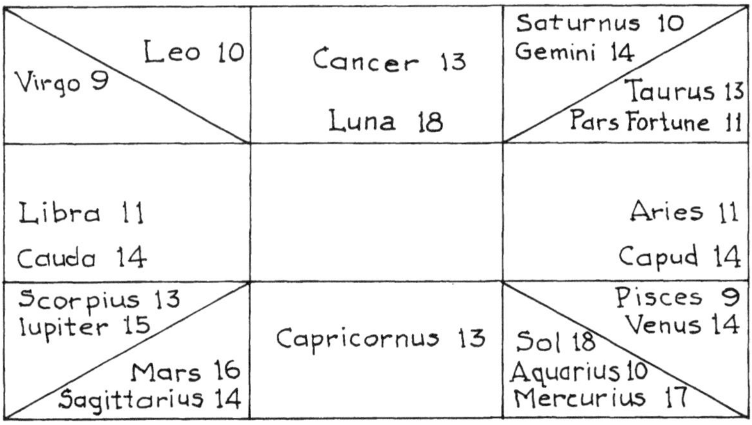

[Horoscope 3.11.] Aspexi in nativitate istius ex dominis triplicitatis Lune, quorum primus fuit Mars, secundus Venus. Et hos inveni cadentes ex angulis. Designat habere malam vitam et amissionem. Cum aspeximus Lunam, et erat in 10ª, et Pars Fortune in Tauro, indicat eum extremo tempore sue etatis habere bonum.

17. 10ª] tauro 18. tauro] domo *written above* 11°

Kitāb al-mawālīd

nativitas diurna

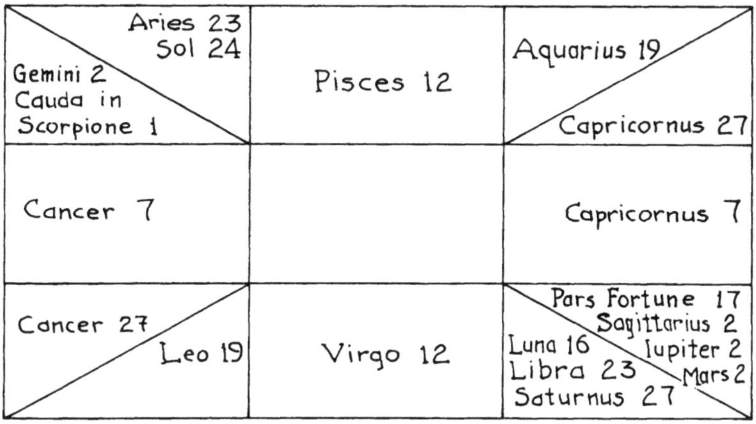

[Horoscope 3.12.] Aspexi ex dominis triplicitatis domus in qua Sol erat, quorum primus erat Sol, Iupiter 2us, 3us Saturnus; aspexique ad Partem Fortune. Indicat eum bonum habere et honorem, et magis in medietate sui temporis. Et hoc apparet in Figura.

17. aparet

Horoscope 3.1. Māshā'allāh has taken this horoscope from the *Pentateuch* of Dorotheus of Sidon (1, 23); and, like all of these horoscopes, it was borrowed from him by his pupil, Abū 'Alī al-Khayyāṭ; for the present I have used only the Latin translation, *Albohali Arabis astrologi antiquissimi ac clarissimi De iudiciis nativitatum*, ed. Ioachimus Heller (Noribergae, 1546), chapter 7.

	Dorotheus	Māshā'allāh	Abū 'Alī	Computation (2 Aug. 43)
♄	♏	♏	♏	♏19
♃	♉	♓	♉	♉18
♂	♒	♒	♒	♏ 9
☉	♌	♌	♌	♌ 7
♀	♌	♌	♌	♌ 6
☿	—	♍	♍	♍ 4
☾	♏	♏	♏	♏11
H	♊	♊ 6	♊	
☊			♋	

Dorotheus' interpretation of the horoscope is as follows: "The birth is nocturnal and, as for his happiness, you find that the first planets are Mars and Venus, because they are the lords of the triplicity to which belongs the sign in which the Moon is; they are both cadent. This man will be needy and poor, cannot get his daily bread, and will find great difficulties; and he will be tortured."

Abū 'Alī says of it: Significatur in ista figura infortunium istius nati a dominis triplicitatis Lunae, quia fuit nativitas nocturna. Erat autem dominus huius triplicitatis primus Mars, secundus Venus, ambo cadentes ab angulis, qui significabant paupertatem et malum esse nati. Quare etiam pauper fuit natus iste nec habuit victum suum nisi post laborem.

It is amusing to catch Māshā'allāh using the first person, as if he had cast this horoscope himself.

Horoscope 3.2. Māshā'allāh has taken this horoscope from some unknown sixth-century Greek source, from which he has also derived horoscopes 3.5 (?), 3.6-3.11, and 3.12 (?).

	Māshā'allāh	Abū 'Alī	Computation (29 Jan. 425)
♄	♏	♏	♏ 19
♃	♋	♋	♋ 4
♂	♏	♏	♏ 29
☉	♒	♒	♒ 6
♀	♑	♑	♐ 25
☿	♒	♒	♒ 1
☾	♐	♐	♐ 8
H	♈ 12	♈	

Abū 'Alī says of it: Significatur fortuna ipsius nati a dominis triplicitatis Solis, quia fuit nativitas diurna. Erat enim dominus triplicitatis huius primus Saturnus, secundus Mercurius, ambo in succedentibus angulorum; Saturnus quidem in succedenti angulo occidentis et Mercurius in succedenti angulo medii coeli, qui significabant prosperitatem ac divitias et multitudinem substantiae. Fuit ergo hic natus in re latissima abundans pecuniis in summo honore ac maxima prosperitate.

Horoscope 3.3. Māshā'allāh has taken this horoscope also from Dorotheus' *Pentateuch* (1, 23), although he has had to guess at the position of Mars, which the Pahlavi version of Dorotheus omitted.

	Dorotheus	Māshā'allāh	Abū 'Alī	Computation (30 March 22)
♄	♓	♓	♓	♒ 29
♃	♋	♋	♋	♋ 10
♂	—	♏	♏	♋ 15
☉	♈	♈	♈	♈ 7
♀	♍	♍	♍	♒ 21
☿	♈	♈	♈	♈ 11
☾	♋	♋	♋	♋ 22
H	♋	♋	♋	

Dorotheus' prediction is: "The birth is diurnal, and the first lord of the triplicity (of the Sun) is the Sun, the second Jupiter; both are in cardines in their exaltations. This child will be thanked by kings, noblemen, and rich people. Saturn is the third lord of the triplicity and is cadent from a cardine; but, because it is in a house of Jupiter aspected by it in trine, he will be praised as kings are praised."

Abū 'Alī has: Haec figura significat fortunam nati a dominis triplicitatis Solis, quia fuit nativitas diurna. Et fuerunt huius triplicitatis dominus primus Sol, secundus Iupiter, ambo in angulis et exaltationibus suis, qui significabant eius sublimitatem et fortunam ac facultatum multitudinem, famam bonam, et laudem maximam apud reges et nobiles. Et Saturnus dominus triplicitatis tertius non impediebat eum quamvis erat cadens ab angulo in domo Iovis et aspiciebant se trino aspectu; ideo significabant quod ipse natus fuerit honoratus et collaudatus apud reges altissimos

Horoscope 3.4. This horoscope Māshā'allāh has taken from Dorotheus (1, 23).

	Dorotheus	Māshā'allāh	Abū 'Alī	Computation (2 April 36)
♄	♍	♍	♍	♌ 28
♃	♍	♍	♍	♍ 16
♂	♒	♒	♒	♓ 5
☉	♈	♈	♈	♈ 11
♀	♉	♉	♉	♉ 27
☿	♓	♓	♓	♉ 1
☾	♏	♏	♏	♏ 17
H	♏	♏	♏	

Dorotheus says of it: "The birth is nocturnal . . . and the lords of the triplicity of the Moon are Mars, Venus, and the Moon; the three of them are in cardines. This child will be a man of great honor and a great chief such that crowns of gold and silver will be placed on him, and he will be praised."

Abū 'Alī predicts: Significatur fortuna ipsius nati a dominis triplicitatis Lunae; fuit enim nativitas nocturna. Dominus primus triplicitatis Lunae fuit Mars, secundus Venus, tertius Luna, omnes in angulis, qui significaverunt prosperitatem, sublimitatem, regnum. Et ideo fuit vir iste magnus, excelsus, et stabilis in prosperitate donec corona aurea fuit posita in caput eius ac argentea cum margaritis et lapidibus preciosis, ac laus eius fuit in ore omnium hominum.

Horoscope 3.5. Māshā'allāh seems to have taken this horoscope also from his unknown sixth-century source, though the position of Jupiter is somewhat doubtful. I offer two possible solutions; that of 9 November 542 appears to me the better.

	Māshā'allāh	Abū 'Alī	Computations (9 Nov. 542)	(2 Nov. 747)
♄	♏	♏	♏15	♏5
♃	♍	♊	♊24	♎3
♂	♌	♌	♌2	♌2
☉	♏	♏	♏14	♏13
♀	♌	♌	♑0	♏8
☿	♏	♏	♏25	♏28
☾	♍	♊	♊1	♍10
H	♋	♋		

Abū 'Alī says of it: Significatur fortuna istius nati a dominis triplicitatis Lunae, quia nativitas fuit nocturna. Eius triplicitatis dominus primus fuit Mercurius et secundus Saturnus, cadentes ab angulis, quorum significatio penuriam et paupertatem portendebat. Sed Iupiter tertius dominus huius triplicitatis erat in succedenti angulo medii coeli, qui temperavit et refrenavit infortunium proveniens ab aliis duobus praedictis. Fuit ergo natus iste mediocris vitae, qualis est religiosorum.

Horoscope 3.6. This is yet another example that Māshā'allāh has taken from his sixth-century source.

	Māshā'allāh	Abū 'Alī	Computation (25 March 434)
♄	♓	♓	♓0
♃	♈	♈	♈14
♂	♍	♍	♌10
☉	♈	♓	♈6

	Māshā'allāh	Abū 'Alī	Computation (25 March 434)
♀	♉	♉	♉ 20
☿	♈	♈	♓ 9
☾	♓	♓	♓ 22
H	♊	♊	

Abū 'Alī's prediction is: Consideravi fortunam istius nati a dominis triplicitatis Solis, quia fuit nativitas diurna; cuius dominus primus erat Sol, secundus Iupiter, tertius Saturnus, omnes in angulis existentes, quorum significatio fortunam et multitudinem divitiarum insinuabat. Et ideo vir iste fuit sublimis ac dives, habens plurimun auri et argenti.

Horoscope 3.7. Here is still another horoscope from the same source.

	Māshā'allāh	Abū 'Alī	Computation (19 Jan. 403)
♄	♒	♒	♒ 4
♃	♍	♍	♍ 7
♂	♑	♑	♒ 15
☉	♒	♒	♒ 0
♀	♐	♐	♑ 5
☿	♒	♒	♒ 12
☾	♊	♊	♊ 9
H	♍	♍	

Abū 'Alī says of it: Significatur fortuna istius nati a dominis triplicitatis Lunae; fuit enim nocturna genesis. Et erat dominus primus triplicitatis Mercurius, secundus Saturnus, ambo cadentes, qui significabant paupertatem et malum esse istius nati. Et ita accidit adeo ut vir iste tenuis fortunae fuerit nullis abundans rebus nisi post sudorem ac laborem. Sed quia Iupiter ac Venus erant in angulis, significaverunt sanitatem corporis eius ac bonam nutritionem et victum a regibus ac principibus multisque amicis.

Horoscope 3.8. Also from the sixth-century source.

Kitāb al-mawālīd

	Māshā'allāh	Abū 'Alī	Computation (18 Oct. 439)
♄	♉11	♉15	♉11
♃	♎21	♎21;12	♎5
♂	♍18	♍18	♍19
☉	♎ 6;8	♎ 8	♎26
♀	♍28	–	♎ 0
☿	♏11	♏11;15	♏14
☾	♌29	♌26	♌29
H	♏ 1	♏ 1	
☊	♉ 8	♉ 8;40	

Abū 'Alī predicts: Haec constellatio significat fortunam istius nati a dominis triplicitatis Solis, quia nativitas fuit diurna; cuius dominus primus fuit Saturnus, secundus Mercurius, tertius Iupiter, quorum quisque erat detrimentum alii. Saturnus enim erat in oppositione Mercurii coniunctus Capiti Draconis ac cadens; Iupiter similiter cadens; et Pars Fortunae cum Marte; et dominus Partis coniunctus Caudae in oppositione Saturni; qui significaverunt laborem et paucitatem bonorum ac perturbationem sensuum, quae sic acciderunt huic nato.

Horoscope 3.9. Yet another taken from the sixth-century source.

	Māshā'allāh	Abū 'Alī	Computation (25 Nov. 464)
♄	♓14	♓15	♓6
♃	♏23	♏22	♏5
♂	♐15	♐19	♐22
☉	♐8	♐9	♐5
♀	♏12	♏13	♏13
☿	♏21	♏21	♐25
☾	♈16	♈16	♈16
H	♊29	♊16	
☊	♑23	♑23	

Abū 'Alī makes the prediction: Significabatur fortuna istius nati a dominis triplicitatis Lunae, quia fuit nativitas nocturna; cuius dominus primus Iupiter, secundus Sol, ambo cadentes ab angulis in signo sextae, qui insinuabant significationem laboris nati ac paucitatem bonorum. Deinde aspexi Partem Fortunae et inveni eam in signo nono, domino suo eam non respiciente, quae etiam laborem et egestatem significabat.

Horoscope 3.10. This horoscope comes originally from the *Thesauri* of Rhetorius of Egypt.* Presumably the unknown sixth-century source included it in his collection, and thence it came to be known to Māshā'allāh; Rhetorius' *Thesauri* was not available in Arabic.

	Rhetorius	Māshā'allāh	Abū 'Alī	Computation (8 Sept. 428)
♄	♐ 14;21	♐ 14	♐ 14	♐ 17
♃	♎ 15;41	♎ 16	♎ 17	♎ 18
♂	♍ 21;6	♍	♍ 21	♍ 24
☉	♍ 14;19	♍ 17	♍ 17	♍ 17
♀	♌ 25;40	♌ 25	♌ 17	♌ 29
☿	♎ 3;37	♎ 4	♎ 5	♎ 8
☾	♓ 3;4	♓ 3	♓ 1	♓ 7
H	♉ 25;16	♉ 11	♉ 21	
☊	♑ 3;41	♑ 8	♑ 8	♑ 6

Rhetorius gives no interpretation, but Abū 'Alī does: Significatur fortuna istius nati a dominis triplicitatis Lunae, quia nativitas fuit nocturna; cuius dominus primus Mars, sub radiis Solis in quadrato aspectu Saturni, qui significavit laborem et anxietatem nati in prima tertia vitae eius. Secundus autem dominus triplicitatis erat Venus, et ipsa domina ascendentis in angulo, orientalis, et augmentata numero, significans prosperitatem et bonum esse nati post laborem.

*See *Catalogus Codicum Astrologorum Graecorum* (Brussels, 1898-1953), vol. 8, part 1, pp. 221-222, and O. Neugebauer and H.B. Van Hoesen, *Greek Horoscopes* (Memoirs, No. 48; American Philosophical Society, Philadelphia, 1959), pp. 138-140, no. L 428.

Decernunt etiam Iupiter et Mercurius, qui sunt in signo sextae descendentes ad quintam, apertae fortunae prosperitatem et bonum statum nati. Atque ita accidit nato.

Horoscope 3.11. This is yet another example from the sixth-century source.

	Māshā'allāh	Abū 'Alī	Computation (7 Feb. 442)
♄	♊ 10	♊ 2	♊ 5
♃	♏ 15	♐ 15	♐ 17
♂	♐ 16	♐ 15	♐ 19
☉	♒ 18	♒ 10	♒ 20
♀	♓ 14	♓ 25	♓ 27
☿	♒ 17	♒ 15	♒ 10
☾	♋ 18	♋ 8;4	♋ 4
H	♎ 11	♎ 2;30	
☊	♈ 14	—	

Abū 'Alī says of it: Haec figura significat fortunam istius nati a dominis triplicitatis Lunae; fuit enim nativitas nocturna. Cuius dominus primus erat Mars, secundus Venus, ambo cadentes in sexta et tertia, qui significabant malum esse nati. Sed Luna, quae erat luminare temporis, erat in medio coeli, ipsa quoque domina triplicitatis postrema, et Pars Fortunae de natura Veneris; significabant pulchritudinem et bonum esse nati in fine vitae suae. Quandocumque enim ceciderint domini triplicitatis luminaris cuius fuerit dominium, aspice Partem Fortunae, quia ipsa significat res maximas et excelsas quando coniuncta fuerit Iovi vel Veneri vel eisdem applicans radiis.

Horoscope 3.12. This last example is the most difficult to date satisfactorily. Māshā'allāh's diagram puts Saturn in the fifth place between Libra 23 and Sagittarius 2, and Jupiter and Mars in the sixth between Sagittarius 2 and Capricorn 7. Abū 'Alī has one interpretation of this situation, but it is undatable; I propose another, which places this horoscope within the series derived from the unknown sixth-century source, though it is still unsatisfactory with regard to Jupiter.

	Māshā'allāh	Abū 'Alī	Computation (16 April 455)
♄	♎27 or ♏27	♎27	♏26
♃	♐ 2 or ♑2	♐17	♒ 0
♂	♐ 2 or ♑2	♐	♐22
☉	♈24	♈24	♈27
♀	–	–	♈17
☿	–	–	♈29
☾	♎16	♎17	♎7
H	♋ 7	♋10;7	
	♏1	♏17	

Abū 'Alī says of it: Significabatur fortuna istius nati a dominis triplicitatis Solis; erat enim nativitas diurna. Et erat Sol dominus triplicitatis primus in exaltatione sua, in medio coeli, applicatus Saturno, nec receptus ab eo; et Iupiter, dominus secundus, et dominus medii coeli fuere cadentes in signo sextae; Saturnus autem dominus tertius in exaltatione sua significabat nato prosperitatem in fine vitae suae. Aspexi porro iterum Partem Fortunae; et inveni eam in signo sextae iunctam Iovi et Lunam in medio coeli a Parte Fortunae, qui significavere fortunam a medio vitae usque ad finem eius.

In conclusion one may state that Māshā'allāh's method of interpreting individual horoscopes on the basis of the lords of the triplicity of the Sun or the Moon is taken from Dorotheus' *Pentateuch* along with three of his examples. The remaining nine come from an unknown Byzantine source of the sixth century which has drawn upon Rhetorius. Māshā'allāh's pupil Abū 'Alī al-Khayyāṭ simply reiterates his master's message.

APPENDIX 4

Additional Horoscopes of Māshā'allāh

Our most prolific source of additional horoscopes is the *Liber Messahalae de receptione*, ed. J. Heller (Noribergae, 1549), in which are found six examples, which have previously been dated by O. Neugebauer. Five of these are from the first half of 791, the last from the end of 794.

Horoscope 4.1.

	Text	*Zīj al-Shāh*	Computation (11 April 791)
Saturn	Aries 10;15	Aries 9;57	Aries 16
Jupiter	Taurus 19;15	Taurus 20;18	Taurus 24
Mars	Gemini 17;30		Gemini 23
Sun	Aries 20;30		Aries 25
Venus	Pisces 5;37		Pisces 11
Mercury	Aries 24;50		Taurus 1
Moon	Taurus 26;25		Gemini 2
Ascendant	Virgo 15		

Horoscope 4.2.

	Text	*Zīj al-Shāh*	Computation (13 Feb. 791)
Saturn	Aries 4;15	Aries 2;24	Aries 9
Jupiter	Taurus 9;13	Taurus 9;58	Taurus 14
Mars	Taurus 15;18		Taurus 22
Sun	Aquarius 24;35		Aquarius 29
Venus	Capricorn 11;39		Capricorn 16
Mercury	Aquarius 2;7		Aquarius 3
Moon	Aries 28;37		Taurus 2
Ascendant	Leo 28		

Horoscope 4.3.

	Text	*Zīj al-Shāh*	Computation (22 March 791)
Saturn	⟨Aries⟩	Aries 6;33	Aries 14
Jupiter	Taurus 14;15	Taurus 16;16	Taurus 20
Mars	Gemini 6;30		Gemini 11
Sun	Aries 1;20		Aries 3
Venus	Aquarius 16;41		Aquarius 17
Mercury	Pisces 19;15		Pisces 17
Moon	Leo 28;58		Virgo 4
Ascendant	Taurus 28		

Horoscope 4.4. This horoscope has previously been published by E. S. Kennedy, "A Horoscope of Messehalla in the Chaucer Equatorium Manuscript," *Speculum 34* (1959), 629-630.

	Text	*Zīj al-Shāh*	Computation (22 May 791)
Saturn	Aries 16	Aries 13;47	Aries 21
Jupiter	Taurus 28;31	Taurus 29;36	Gemini 4
Mars	Cancer 11;58		Cancer 19
Sun	Gemini 0;5		Gemini 4
Venus	Aries 22;13		Aries 27
Mercury	Gemini 13;16 retr.		Gemini 16 retr.
Moon	Sagittarius 8;18		Sagittarius 8
Ascendant	Gemini 5		

Horoscope 4.5.

	Text	*Zīj al-Shāh*	Computation (20 April 791)
Saturn	Aries 11	Aries 10;8	Aries 17
Jupiter	Taurus 21;10	Taurus 21;0	Taurus 26
Mars	Gemini 22;26		Gemini 29
Sun	Aries 29;10		Taurus 3
Venus	Pisces 17;9		Pisces 21
Mercury	Taurus 10;15		Taurus 20
Moon	Libra 4;22		Virgo 28
Ascendant	Cancer 14		

Horoscope 4.6.

	Text	*Zīj al-Shāh*	Computation (30 Nov. 794)
Saturn	Taurus 29;54 retr.	Taurus 27;58	Gemini 4 retr.
Jupiter	Virgo 20;50	Virgo 21;25	Virgo 27
Mars	Cancer 25;10 retr.		Leo 3 retr.
Sun	Sagittarius 8;15		Sagittarius 13
Venus	Sagittarius 2;31		Sagittarius 6
Mercury	Scorpio 29;18		Scorpio 27
Moon	Capricorn 22;18		Capricorn 21
Ascendant	Sagittarius 9		

Considering that the text is a sixteenth-century edition of a translation of the original Arabic, the agreement between the longitudes given in the text and those computed by means of the reconstructed *Zīj al-Shāh* is rather impressive, but not sufficient to substantiate a claim that Māshā'allāh was using that zīj in 791.

The second most prolific source – producing three horoscopes – is Vaticanus Graecus 1056, a fourteenth-century copy of a late twelfth-century collection of *astrologica.*[1] The first of these three horoscopes comes from Māshā'allāh's sixth-century Byzantine source; the other two and the fourth given here do not seem to have been computed by means of the *Zīj al-Shāh.*

The first Māshā'allāh horoscope Horoscope 4.7 is on f 48v (book 1, chapter 22):

1. For the date of this manuscript see *Dumbarton Oaks Papers* 18 (1964), 138-139.

Περὶ τοῦ γνῶναι τὴν ἐρώτησιν καθὼς καὶ αὖθις ὁ Μασάλα φησίν, ποιήσας καὶ ὑπόδειγμα.

[f. 49] Περὶ τοῦ γνῶναι τὴν ἐρώτησιν κατὰ τὸ προκείμενον ὑπόδειγμα ὁ Μασάλλα φησὶν ὅτι, ἐπειδὴ τὸ ὡροσκοποῦν ζῴδιόν ἐστι Ταῦρος, εὑρέθη δὲ ὁ Ζεὺς ἐκεῖσε, ἡ δὲ κυρία τοῦ ὡροσκόπου Ἀφροδίτη ἦν ἐν Καρκίνῳ τῷ τρίτῳ τόπῳ ἐν τῷ οἰκείῳ αὐτῆς[2] τριγώνῳ, ἡ δὲ Σελήνη ἦν ἐν Ζυγῷ μὴ οὖσα ἐν οἰκείῳ τόπῳ, ὁ δὲ Ἥλιος ἐν Τοξότῃ μηδ' αὐτὸς ὑπάρχων ἐν οἰκείῳ τόπῳ, ὁ δὲ κύριος τοῦ κλήρου τῆς τύχης ἦν ὁ Κρόνος καὶ ἔστιν ἐν τῷ ὑπογείῳ τετάρτῳ τόπῳ, οὐχ ἕτερός τις προεκρίθη ἀστὴρ σημειωτικὸς τῆς τοιαύτης ὑποθέσεως ἢ ἡ Ἀφροδίτη διὰ τὸ εἶναι αὐτὴν ἐν τῷ οἰκείῳ τριγώνῳ. καὶ ἐπειδὴ ἡ οἰκοδέσποινα αὐτῆς Σελήνη ὑπάρχει ἐν Ζυγῷ τῷ τετάρτῳ τόπῳ ἀπὸ τοῦ[3] οἴκου αὐτῆς, ἔστι δὲ καὶ ἡ Σελήνη σημειωτικὴ τῶν μητέρων, ἐπέγνωμεν εἶναι τὴν ἐρώτησιν περὶ μητρὸς καὶ ὅτι καὶ ἀρρωστεῖ ἡ μήτηρ διὰ τὸ εἶναι τὴν Σελήνην ἐν τῷ ἀπὸ τοῦ ὡροσκόπου ς' τόπῳ. ὁρῶντες δ' αὖθις ἀπορρεῦσαν[4] τὴν Σελήνην ἀπὸ τῆς ἑξαγώνου συναφῆς τοῦ Κρόνου, ἐπέγνωμεν εἶναι τὴν τοιαύτην νοσηλείαν ἀπὸ ξηρότητος καὶ ψυχρότητος κατὰ τὴν τοῦ Κρόνου κρᾶσιν. ἐπεὶ δὲ εἴδομεν εἶναι τὸν Κρόνον ἐν Λέοντι, διεγνώσθη ἡμῖν νοσεῖν τὸ κυριευόμενον μέρος τοῦ Λέοντος ἤγουν τὸν στόμαχον. ταῦτα σκεψάμενοι, ἀναγκαῖόν ἐστι μαθεῖν καὶ εἰπεῖν ἡμᾶς καὶ τὸ τῆς νόσου ταύτης ἀποτέλεσμα. καὶ ἐπειδὴ ὁρῶμεν τὴν Σελήνην ἀπερχομένην πρὸς τετράγωνον συναφὴν τῆς ἀγαθωτάτης Ἀφροδίτης, λέγομεν ἀπαλλαγῆναι τὸν νοσοῦντα τῆς νόσου διὰ τὴν τῆς Ἀφροδίτης ἀγαθότητα. τὸ δὲ διὰ πόσου χρόνου τοῦτο γενήσεται· ἐπειδὴ πέντε μοίρας λείπεται ἡ Σελήνη πρὸς τὸ τετραγωνίσαι τὴν Ἀφροδίτην, ἔστι δὲ ἡ Σελήνη ἐν τροπικῷ ζῳδίῳ, λέγομεν ἀπαλλαγῆναι τῆς νόσου ὁ νοσῶν διὰ πέντε ἡμερῶν. εἰ γὰρ ἦν ἡ Σελήνη ἐν δισώμῳ ζῳδίῳ, εἴπομεν ἂν μῆνας ε̅. εἰ δὲ ἐν στερεῷ ζῳδίῳ, ἐνιαυτοὺς πέντε. εἰ δὲ καὶ ἀπήρχετο[5] ἡ Σελήνη πρὸς συναφὴν τοῦ Ἄρεος, ἔμελλε θανεῖν ὁ νοσῶν διὰ τοῦτο συνεχῶς.

Εἰ δὲ ἦν ἡ Ἀφροδίτη ἔνθα αὐτὴ ἡ Σελήνη ἐστίν, πάλιν εἴχομεν εἰπεῖν ὅτι περὶ μητρός ἐστιν ἡ ἐρώτησις· εἰ δὲ ἦν ὁ Κρόνος, εἴπομεν περὶ πορνείας· εἰ δὲ ἦν ὁ Ἄρης, εἴπομεν περὶ φυγάδος· εἰ δὲ ἦν Ἥλιος, εἴπομεν ὅτι περὶ μεγίστου προσώπου ἐστὶν ἡ ἐρώτησις. εἰ δὲ ἦν ἡ Ἀφροδίτη ἐν τῷ ἑβδόμῳ, εἴπομεν περὶ γαμικοῦ συναλλάγματος εἶναι τὴν ἐρώτησιν· εἰ δὲ ἦν ὁ Ἑρμῆς, περὶ γραμμάτων καὶ λοιπῶν ὁμοίων. εἰ δὲ ἦν ὁ Ζεὺς ἐν τῷ θ' τόπῳ, εἴχομεν εἰπεῖν ὅτι περὶ ὀνείρων ἐστὶν ἡ ἐρώτησις. οὕτως οὖν σκέπτου περὶ τῆς ἐρωτήσεως· καὶ πρόσχες μετὰ ἀκριβείας, καὶ μὴ [f. 49v] ἀμελήσῃς τι πρὸ τοῦ ἀποτελέσαι μήτε οὖν καταφρονῇς ταῖς ἀπορροίαις τῶν ἀστέρων, ταῖς δορυφορίαις, ταῖς συναφαῖς, καὶ ταῖς ὑποδοχαῖς αἷς ποιεῖται ὁ εἷς ἀστὴρ πρὸς τὸν ἕτερον· πρὸς δὲ καὶ τὰς δυνάμεις αὐτῶν καὶ τὰς κυβερνήσεις καὶ οὐδέν τί σοι ἐπιλήσεται.

2. αὐτῆς MS. 3. τῆς MS. 4. ἀπορρεύσασαν MS. 5. ἀπέρχετο MS.

The dating of this horoscope is indeed difficult – especially as Venus has an elongation of six signs from the Sun! The most reasonable solution seems to be to take the last sentence in the first paragraph as indicative of the original situation in the horoscope, in which Mars, not Venus, would have been in Cancer. The recovery of the querist's mother led the astrologer to substitute the benefic for the malefic planet in the opposite sign. If this interpretation is accepted, the date of the horoscope is 30 November 446.

	Text	Computation (30 Nov. 446)
Saturn	Leo	Leo 21
Jupiter	Taurus	Taurus 7
Mars	Cancer	Cancer 11
Sun	Sagittarius	Sagittarius 17
Venus		Capricorn 19
Mercury		Scorpio 18
Moon	Libra	Libra 6
Ascendant	Taurus 10	

The second Māshā'allāh horoscope (4.8) in this manuscript is in the upper margin and right-hand margin of f 86.

Τὸν Μασάλα ἐρώτησεν. ὁ Μασάλα τὸ τοιοῦτον ἐξέθετο θεμάτιον. καὶ τὴν ἀπόφασιν ταύτην πεποίηκεν: ἔλαβον πρόσωπον τοῦ ἐρωτήσαντος τὸν ὡροσκόπον, τὸν κύριον αὐτοῦ, καὶ τὸν Κρόνον ἀφ' οὗ ἡ Σελήνη ἀπέρρευσεν· πρόσωπον δὲ τοῦ ἐναντίου τὸ δῦνον, τὸν κύριον αὐτοῦ, καὶ τὴν Ἀφροδίτην τὴν παρὰ τῆς Σελήνης συναπτομένην. λοιπόν· οἱ μὲν τοῦ ἐρωτήσαντός εἰσιν ἀστέρες ὑψηλοί, τοῦ δὲ ἐναντίου χαμηλοί. εὑρέθη γοῦν ὁ Ζεὺς ἐν τῷ ς' τόπῳ ὑποποδίζων, τοῦ Ἄρεως ἀπερχομένου διαμετρῆσαι αὐτὸν ἀπὸ τοῦ ιβ' οἴκου· ὁ δὲ Κρόνος εὑρέθη μὲν ἐν τῷ ιβ' ἐν ᾧ δὲ ...[6] χαίρει· ὁ δέ γε Ἑρμῆς καὶ ἡ Ἀφροδίτη ἐν τῷ ι' τόπῳ εὑρέθησαν, οἱ δὲ διὰ τοῦτο δηλοῦσι δύναμιν τὸν ἐχθρὸν ἔχειν. ὅτε ἐστὶν ὁ Ἑρμῆς καὶ ἡ Ἀφροδίτη ἐν τῷ Ζυγῷ, εἰσὶν ἀκάκωτοι· ὅτε δὲ ἐν τῷ Σκορπίῳ εἰσὶ θάτεροι– ζήτει ἐκεῖθεν– (in upper margin) <ἐ>κ τῶν κακοποιῶν...[6]

6. The writing has disappeared along with the margin.

The date of this second horoscope falls within Māshā'allāh's lifetime.

	Text	*Zīj al-Shāh*	Computation (12 Oct. 778)
Saturn	Scorpio 19	Scorpio 12;33	Scorpio 18
Jupiter	Taurus 12 retr.	Taurus 8;14	Taurus 13
Mars	Scorpio 5		Scorpio 8
Sun	Libra 15;16		Libra 23
Venus	Virgo 30		Libra 2
Mercury	Libra 2		Libra 4
Moon	Taurus 20		Taurus 15
Ascendant	Sagittarius 21		

The third Māshā'allāh horoscope (4.9) in this manuscript is found in the lower left-hand margin and in the bottom margin of f 95v.

Τὸ *τοιοῦτον θεμάτιόν ἐστιν ἐρωτήσεως γενομένης παρά τινος πρὸς τὸν* Μασάλα *περὶ παραμονῆς καταρχῆς ἢ ἐξελεύσεως ἀπὸ τῆς ἀρχῆς αὐτοῦ.* Ἐσκεψάμην *τὴν τοιαύτην ἐρώτησιν· καὶ εὗρον τὸν μὲν κύριον τοῦ ὡροσκόπου ἐν τῷ ια' τόπῳ, τὴν δὲ* Σελήνην *ἐν τῷ* Λέοντι *τῷ ε' τόπῳ μὴ ὁρῶσαν μὲν τὸν οἰκοδεσπότην αὐτῆς, διαμετροῦσαν δὲ τὸν* Ἄρην. *ὁ δὲ τριγωνοκράτωρ τοῦ ὡροσκόπου καὶ ὑψοκράτωρ* Ἥλιος *εὑρέθη ἐν τῷ μεσουρανήματι ἐν ζῳδίῳ τροπικῷ, οἰκοδεσπότης ὢν καὶ τῆς* Σελήνης. *ὁ δὲ οἰκοδεσπότης τοῦ μεσουρανήματος* Κρόνος *εὑρέθη ἐν τῷ τῆς ἀποβάσεως τόπῳ ζῳδίῳ τροπικῷ ὑποποδισμένος. διὰ γοῦν τὸ εἶναι τὰ ζῴδια τοῦ τε μεσουρανήματος καὶ τοῦ δ' τόπου τροπικὰ καὶ τὸν κύριον τοῦ μεσουρανήματος* Κρόνον *ἐν ζῳδίῳ τροπικῷ ὑποποδισμένον δηλοῖ τὸ ἐξωσθῆναι τὸν ἀρχηγὸν ἀπὸ τῆς ἀρχῆς αὐτοῦ· ἀποτελέσθη οὕτω κατὰ τὴν τοῦ τοιούτου σοφοῦ κρίσιν.*

Appendix 4

The same horoscope is discussed (4.10) in the *Jāmi*ᶜ *al-kitāb* of Abū Yūsuf Yaᶜqūb ibn ᶜAlī al-Qaṣrānī (*ca.* 890), but without reference to Māshā'allāh. I edit this from two manuscripts: B is Berlin 5877 p. 487, H is Hamidiye 821 ff 304v – 305.

مسئلة اخرى عن سلطان وبقائه.

وجدت رب الطالع فى بيت الرجاء والقمر فى الخامس ساقطاً عن رب بيته يتصلاً بالمريخ من مقابلة. فلمكان الشمس صاحب مثلثة الطالع فى وسط السماء دل ان المسئلة عن سلطان. ولانقلاب برج الطالع والعاقبة وكون وسط السماء منقلباً وربه راجعاً فى بيت العاقبة دل على زواله عن سلطانه. ولسقوط رب بيت صاحب الطالع عن مناظرته وقبوله دل ان صاحب هذا السلطان مذموم فى اهل عمله جائر شديد بجوره. ولاتصال القمر رب العاقبة بالمريخ من مقابلة دل على توجه صاحب المسئلة الى ارض كثيرة الماء والندى. فحسبت الوقت فى ذلك ومعرفته من الشمس (التى كانت اقوى ادلة) الى الطالع. فكانت تسعة وستين درجة. فدل على رجوعه من ذلك الحرب بعد هذه العدة من الايام ويحترق رب الطالع بعد ذلك

بشهر وهو حيث ينتهي ذهاب سلطانه
فاردت ان اعرف من ياتي بعده على
سلطانه. فطرحت الستين من التسعة والستين.
ثم زدت عليه سهم العمل الثاني وكان الدليل
في ذلك الزهرة وكانت غربية مقبولة. فقلت
ما ياتي من بعده رجل من العرب من
ناحية المغرب غريب ليس من اهل المصر

1 وثباته H | 4 فلان B | 5 in marg. H | 5 في om. B | 6 عن] لصاحب H | 14 H adds حرب after ارض | 14 الحرب والندا B | 15 H adds لمعرفة after فحسبت | 15 H omits ومعرفته في ذلك | 16 الى om. H | 17 H has سبع وستون in the text, تسعة وستون in marg. | 17 H adds ان after على | 18 ذلك om. B | 19 الايام] ذلك B | 20 يبين فهو H | 21/25 instead of ان اعرف —— من بعده H has علم ما ياتي بعده على سلطانه. فطرحت التسيير من الشعاع والتسيير | 25 العرب] المغرب H.

This third horoscope was also computed during Māshā'allāh's lifetime.

	Byzantine text	al-Qaṣrāni	Computation (10 Jan. 768)
Saturn	Cancer 5;26	Cancer 26;22	Cancer 8
Jupiter	Gemini 1;39	Gemini 1;30	Gemini 6
Mars	Aquarius 22	Aquarius 22	Aquarius 26
Sun	Capricorn 17;0	Capricorn 22	Capricorn 24
Venus	Aquarius 12	Aquarius 12	Aquarius 14
Mercury	Sagittarius 29	Sagittarius 29	Capricorn 5
Moon	Leo 18;10	Leo 18	Leo 13
Ascendant	Aries 3	Aries 3;18	
Lunar node	Libra 10	Libra 20	

	Zīj al-Shāh
Saturn	Cancer 2;55
Jupiter	Gemini 0;55

The final horoscope is found in Māshā'allāh's *Kitāb al-as*^{c}*ār*; I have used Escorial Ar. 938, in which it appears on ff 69v-70 and again on f 70.

	Text	*Zīj al-Shāh*	Computation (24 June 773)
Saturn	Virgo 4	Virgo 11;7	Virgo 14
Jupiter	Scorpio 12[7]	Scorpio 6;3	Scorpio 11
Mars	Virgo 11[8]		Pisces 18 (opposition)
Sun	Cancer 10		Cancer 6
Venus	Capricorn 23 (sic!)		Gemini 3
Mercury	Cancer 15		Cancer 10
Moon	Cancer 10		Cancer 4
Ascendant	Gemini 8		

7. 9 on ff 69v-70.

8. 8 on ff 69v-70.

APPENDIX 5

A Note on the Flood Date, and Five Horoscopes from an Erfurt MS

On f 36v of MS Q° 355 of the Bibliotheca Amploniana in Erfurt[1] is an interesting note written in an extremely difficult and much faded hand. I present here an attempt to decipher the beginning of this note, which particularly pertains to Māshā'allāh's date for the conjunction indicating the Flood; both capitalization and punctuation are mine.

> Nota pro (?) Albumasar libro De coniunctionibus planetarum tractatu primo differentia prima: inter coniunctionem significantem sectam Arabum que fuit in Scorpione et coniunctionem significantem diluvium que fuit in Virgine fuerunt 3950 anni.[2] Quid sit (?) deinde exclusis ambobus (?) annis communalibus ? Prima enim fuit ante Christum 3382° anno, alia fuit anno Christi 570°, tam media quam vera. Verum (?) exclusis

1. I take this opportunity to express my gratitude to Dr. E. Boer of Berlin for making possible both my brief personal inspection of this manuscript and my longer study of ff. 36v-37 on microfilm. D.P.
2. *Albumasar De magnis coniunctionibus annorum revolutionibus ac eorum profectionibus* (Venetiis, 1515), tractatus 1, differentia 1 (p. 7): coniunctio que significavit diluvium fuit ante coniunctionem significantem sectam Arabum per 3950 annos, et praefuit illi orbi Saturnus cum signo Cancri. The whole question of Abū Ma'shar's use of Māshā'allāh's chronological scheme is discussed in detail in D. Pingree, *The Thousands of Abū Ma'shar* (Studies of the Warburg Institute, 30; London, 1968), pp. 40-42 and 60 n. 1.

> utroque annis communalibus,[3] si addas 569 perfectos et 3381 perfectos, exibunt 3950. Et ad completationem (?) orbis 11 deest[4] 10 anni solares; quibus convolutis verificatur deinde an corrigitur idem. Idem dicit Omar Libro nativitatum capitulo ultimo;[5] idem ex textu. Etiam narrat Belenus Bentemiz[6] et alii extra eum quod inter creationem Ade et noctem diei saevioris (?) in quo fuit diluvium fuerunt anni 2226 et mensis unus et 23 dies et 4 hore.

On f 37 of this same manuscript are six horoscopic diagrams, of which five have been filled in. The first three relate to the vernal equinoxes of the years in which the conjunction indicating the Flood occurred (–3380), in which the Flood itself occurred (–3101, the epoch of the present Kaliyuga), and in which the second cycle (*orbis*) of 360 years began (–3020); all three are dated from an epoch, –3388, whose significance escapes me. Of the last two horoscopes, the first is for the vernal equinox of A. D. 1351, the second for the immediately preceding conjunction of the Sun and Moon. The whole series seems to have been computed independently by the author of the note on f 36v in about the year 1351.

Horoscope 5.1. Written above is: Iste annus fuit ante Christum 3381, et fuit inicium primi orbis. In the center is written: Introitus anni 8 scilicet 19d. 21h. 58m. Februarii. Iste annus fuit ante Christum 3381, et fuit inicium primi orbis; et pervenit directio ad primum gradum Scorpionis; et praefuit orbi Saturnus cum signo Cancri.

	Text (19 Feb. –3380)	Māshā'allāh (11 Feb. –3380)
Saturn	Virgo 23;9 retr.	Scorpio 1;56 retr.

3. annorum communalium MS.
4. dest MS.
5. Though there are several references to the *revolutiones annorum mundi*, there is no discussion of chronological problems in *De nativitatibus secundum Omar*, ed. N. Pruckner in *Iulii Firmici Materni . . . Astronomicωn Libri VIII* (Basileae, 1551), pt. 2, pp. 118-141; but for the attribution of a similar statement to Omar (presumably 'Umar ibn al-Farrukhān al-Ṭabarī), see Appendix 6.
6. Belenus is usually identified with (pseudo-)Apollonius of Tyana.

Jupiter	Libra 3;51 retr.	Libra 12;39 retr.
Mars	Scorpio 26;29	Libra 27;17 retr.
Sun		Aries 0;1
Venus	Aquarius 22;11	Pisces 26;16
Mercury	Pisces 8;44	Pisces 3;24
Moon	Gemini 25;52	Taurus 23;13
Ascendant	Gemini 25	Sagittarius 19
Node	Scorpio 17;47	Cancer 8

Horoscope 5.2. In the center is written: Introitus anni 287 scilicet 17d. 0h. 21m. Februarii. Iste est annus diluvii secundum omnes astrologos. –3388 plus 287, of course, equals –3101.

Saturn	Pisces 5;53
Jupiter	Aries 16;58
Mars	Pisces 20;50
Sun	Aries 0;0,1
Venus	Aries 6;57
Mercury	Aries 26;19
Moon	Aries 13;42 (?) or 2;13,42 (?)
Ascendant	Cancer 29
Node	Virgo 21;51

Horoscope 5.3. In the center is written: Introitus anni 368 scilicet 16d. 16h. 56m. Februarii. Iste annus fuit ante Christum 3021 annis, et incipit 2 orbis cui praefuit Iupiter cum signo Leonis.

Saturn	Scorpio 19;39 retr.
Jupiter	Aquarius 19;8
Mars	Aries 3;35,2
Sun	Aries 0;0,1
Venus	Aquarius 22;49
Mercury	Aries 21;5,25

Moon	Capricorn 20;50
Ascendant	Aquarius 25
Node	Gemini 15;9

Horoscope 5.4. This is very difficult indeed to read. In the center all that I can decipher is: Figura revolutionis anni mundi anno Christi 1351. This is followed by a series of numbers.

	Text	Computation (12 March 1351)
Saturn	Aries 26;52	Aries 25
Jupiter	Leo 11;- (?)	Leo 11
Mars	Aries 10;50	Aries 10
Sun	Aries 0;0,1	Aries 0
Venus	Pisces 16;11	Pisces 15
Mercury	Pisces 21;7	Pisces 20
Moon	Virgo 16;4	Virgo 17
Ascendant	Leo 25;52	
Node	Gemini 12;15	

Horoscope 5.5. In the center I can read only: Figura coniunctionis Solis et Lune anno 1351.

	Text	Computation (26 February 1351)
Saturn	Aries 25	Aries 23
Jupiter	Leo 13	Leo 13
Mars	Aries 15 (?)	Aries 0
Sun	Pisces 17	Pisces 16
Venus	Aquarius 28;40	Aquarius 28
Mercury	Aquarius 25	Aquarius 25
Moon	Pisces 17	Pisces 16
Ascendant	Leo 21	
Node	Gemini 15	

APPENDIX 6

Another Note on the Flood Date

Lynn Thorndike has published, with an English translation,[1] a short anonymous text written in 1428 in which another attempt – not so successful this time – is made to date the conjunction indicating the Flood. I quote paragraphs 5 and 6, which contain this computation:

Media autem coniunctio Saturni et Iovis fuit ante annum diluvii 280 annis 33 diebus. Si igitur tunc fuit inicium orbis, ut volunt astronomi, sequitur quod Christus natus erat anno 142 decimi orbis. Aomar autem ponit initium orbis ante annum diluvii 279 annos ex quo sequi videtur quod inicium orbis erat ante[2] coniunctionem illam magnam que fuit secundum medium motum in septimo gradu Cancri ad meridiem Cantabrius. Preterea coniunctio illa magna de qua loquitur Ovidius tertio de Vetula fuit secundum medium motum in tertio gradu Arietis ante annum nativitatis Christi per 5 annos 320 dies 12 horas 37 minuta 48 secunda. Sic igitur Christus natus est 42 anno Augusti Ceseris cuius oppositum tamen scribitur in illo libro. Et nota quod eorum media coniunctio est in omnibus 19 annis 314 diebus 15 horis, unde ante illam coniunctionem que immediate precessit

1. L. Thorndike, "Astronomical and Chronological Calculations at Newminster in 1428," *Annals of Science* 7 (1951), 275-283. Both the edition and the translation are rather unreliable; but, lacking a film of the manuscript, I have refrained from making any but the most obvious of corrections.
2. Supplied by Thorndike.

Christi nativitatem fuit alia coniunctio ad 3376 annos et 31 dies, igitur illa fuit ante annum incarnacionis per 3381 annos 351 dies et 18 horis. Quare sequitur tempus positum ab Aomar a veritate deficere in uno anno et amplius. Quod hoc sit verum probatur, nam prior fuit ad annos 3381 et fere 352 dies ante nativitatem Christi, de quibus si subtrahantur anni menses et dies diluvii, remanent 280 anni et 33 dies. Unde patet propositum, est enim precise tempus a coniunctione ad coniunctionem 19 anni 314 dies 14 horae 59 minuta 38 secunda 29 tertia. Et si multiplicaveris tempus inter duas coniunctiones Saturni et Iovis per 170, fient 3376 anni 31 dies 23 hore et 23 minuta quo tempore necesse est eos iterato coniungi. Rursus si 3376 annos[2] 31 dies 6 horas diviseris per 170, exibit tempus inter duas coniunctiones videlicet anni 19 dies 313 hore 20 minuta 57 secunda 46.

Alfonsus incepit annum diluvii a 16 die Februarii. Additis igitur 46 diebus[2] et 6 horis ad differentiam diluvii et incarnacionis positam ab Alfonso, resultant precise anni 3102 qui sunt 9 quarta, 14 tertia, 43 secunda, 26 prima, quibus additis 279 annis 352 diebus cum 18 horis fiunt anni ante incarnacionem Ihesu Christi 3381, 352 dies, 18 hore, quo tempore fuit illa coniunctio duorum superiorum planetarum secundum medios cursus in septimo gradu Cancri fere in meridie 18 diei Ianuarii quo anno fuit inicium magni orbis secundum Aomer, et prefuit orbi Saturnus cum signo Cancri. Fuit autem primus dies annorum diluvii feria vii scilicet primus dies Ianuarii et tertius post bisextum, igitur anno coniunctionis fuit primus dies Ianuarii feria vii et sic dies coniunctionis erat feria tertia. Super c. litteram erat quorum ille annus iterum tertius post bisextum que est ratio certissima quod media coniunctio Saturni et Iovis que fuit decima ante diluvium processit annum incarnacionis 3381 annis 352 diebus 18 horis 39 minutis. Erat igitur 13[a] dies Ianuarii hora fere 6[a] post meridiem et erat anno illo b littera dominicalis, annus quoque tercius post bisextum et erat tunc inicium orbis ut dicit Aomer libro de nativitatibus et prefuit orbi Saturnus cum signo Cancri, ut ipse dicit, et erat tunc Saturnus et Iubiter coniuncti secundum medium cursum in septimo gradu Cancri uterque retrogradus.

S^a	G^a	M^a	2^a	3^a	
3	6	49	30		locus medius Saturni
3	6	49	26		locus medius lovis
5	19	59	16		argumentum medium
8	26	48	32		medius motus Solis
3	15	0	24		verus motus Saturni retro.
3	12	39	36		verus motus Iovis retro.
0	2	20	48		differencia.

GENERAL INDEX

Note that astrological terms are normally cited only as they occur in the translation; cross-references to the appropriate places in the commentary can then be easily made. The more common terms, such as the names of the planets and of the zodiacal signs, are omitted. In arranging Arabic names in alphabetical order the article *al-* has been ignored.

INDEX OF HOROSCOPES

ARABIC INDEX

The words listed below are astrological and astronomical terms, referred to by folio and line of the text. No attempt has been made to note every occurrence of a word; where the usage is standard and unequivocal, two or three references suffice. The ordering is strictly alphabetical, not in the order of the Semitic roots, except that in the listing *al-* and *abū* have been ignored. Vowel marks do not appear in the text, and they have not been inserted here. As in the text, dots over the *ta' marbūṭa* have usually been left off. If a plural occurs in the text, it is listed, generally following the singular, but broken plurals are also listed individually.

Bei Fragen zur Produktsicherheit wenden Sie sich bitte an:
If you have any questions regarding product safety,
please contact:

Walter de Gruyter GmbH
Genthiner Straße 13
10785 Berlin
productsafety@degruyterbrill.com